Physics and the
Ultimate Significance
of Time

Physics and the Ultimate Significance of Time

BOHM, PRIGOGINE, AND PROCESS PHILOSOPHY

Edited by *David Ray Griffin*

State University of New York Press

"Bohm and Whitehead on Wholeness, Freedom, Causality, and Time," by David Ray Griffin, appeared previously in *Zygon,* Volume 20, number 2, June 1985 (©1985 by the Joint Publication Board of *Zygon,* ISSN 0044–5614).

Published by
State University of New York Press, Albany

© 1986 State University of New York

For information, address State University of New York
Press, State University Plaza, Albany, N.Y., 12246

Library of Congress Cataloging in Publication Data

Main entry under title: Griffin

Physics and the ultimate significance of time

Papers based on a conference held in March 1984 under
the auspices of the Center for Process Studies, Claremont,
California.
Includes index.
1. Space and time—Congresses. 2. Physics—Philosophy
—Congresses. 3. Process philosophy—Congresses.
4. Bohm, David. 5. Prigogine, I. (Ilya) I. Griffin,
David Ray, 1939-   . II. Claremont Center for Process
Studies.
QC173.59.S65P49   1985   530.1'1   85–2782
ISBN 0–88706–113–3
ISBN 0–88706–115–X (pbk.)

Contents

Preface

The notion that physics is in some fundamental sense timeless has been widely accepted. Of course, time in the sense of a measured interval (symbolized by t) occurs in physical formulae. But time as we experience it, with its incessant becoming, its irreversibility, and its absolute distinctions between past, present, and future, is said to be irrelevant to the thought world of physics. Some thinkers go so far as to claim that this absence proves that time in this sense is an illusion. In any case, insofar as the timelessness of physics is accepted, and insofar as physics is taken to be the fundamental natural science, an unbridgeable chasm seems to exist between the world of the natural sciences and the world of the humanities—or, simply, the world of complete human experience with all its richness and complexity, all its hopes and fears. For in the latter nothing is more fundamental than the reality of temporal process. It is presupposed by our every activity, our every passion.

For the physical fundamentalist, who says "if it's not in physics it's not real," the implication is that all those activities and emotions that presuppose temporality are folly, since they presuppose an illusion. Those who cannot accept this reductionism are forced into some form of ultimate dualism, according to which timeless entities and temporal entities are equally real; they are then left to ponder the mystery of how these two kinds of things can combine to form an integrated universe.

Reductionism and dualism, with all their notorious paradoxes, could both be avoided if the fundamental premise of the timelessness of physics were challenged. There are movements in our day in this direction. In March 1984 a conference was held in Claremont, California, under the auspices of the Center for Process Studies, which brought together representatives of three types of thought that are exploring this third possibility: physicist David Bohm, Nobel Prize–winning physicist-chemist Ilya Prigogine, and various advocates of the school of process philosophy inspired primarily by Alfred North Whitehead (with Henri Bergson and others in the background), including physicist Henry Stapp.

The present volume grew out of that conference. Since the conference, all the authors of papers and responses have had time to revise their comments in the light of the discussions. Also, several additional responses have been added, including direct comments by Bohm, Prigogine, and Stapp, about one or more of the others' work. Finally, I wrote an introduction to put the present three-sided conversation about the relation between physics and time within the context of the ongoing discussion of the past 100 years. I take this opportunity also to give public thanks to my secretary, Sue Klopfenstein, who regularly and cheerfully transmutes the illegible into the legible.

I want to record my gratitude to my hosts, Jan van Bragt and James Heisig, and the rest of the staff at the Nanzan Institute for Religion and Culture in Nagoya, Japan, where an air-conditioned room was provided in August of 1984—that hottest of all summers in recorded Japanese history—, making it possible for me to get the introductory chapter written. Without that room, the only opportunity during the year for doing that writing would have passed— which demonstrates again that space and time are inseparable.

I give my sincere thanks to all the contributors to this volume for getting their papers in on time (not to be presupposed even for a book on time!), and especially to Milič Čapek, Andy Bjelland, John Cobb, Tim Eastman, Fred Ferré, Steve Rosen, and Bob Russell for their helpful feedback on a first draft of my introduction. In this regard I give special thanks to Isabelle Stengers, coauthor of the book upon which I mainly relied for understanding Prigogine's thought.

This book would not have been possible without the ongoing support given to the Center for Process Studies by the School of Theology at Claremont and by Claremont University Center and Graduate School; the latter institution is also to be thanked for the use of its Albrecht Auditorium, a room that makes it possible for a conference to be a delight.

Finally, I am grateful to William Eastman, editor, and Peggy Gifford, production editor, of the State University of New York Press, for their roles in getting this volume published.

The technical artwork for the figures was done by Mary K. Siegel of Santa Barbara, California. The index was prepared by John Sweeney.

In any attempt to bridge the domains of experience belonging to the spiritual and physical sides of our nature, time occupies the key position.

—Arthur Eddington (1968, 91)

The history of natural philosophy is characterized by the interplay of two rival philosophies of time—one aiming at its "elimination" and the other based on the belief that it is fundamental and irreducible.

—G. J. Whitrow (1980, 370)

Time, the devouring tyrant.

—Traditional expression

Time, the refreshing river.

—W. H. Auden (cited by Joseph Needham, 1943)

It is a characteristic of thought in physics . . . that it endeavours in principle to make do with "space-like" concepts alone.

—Albert Einstein (1954, 141)

Time and space express the universe as including the essence of transition and the success of achievement. The transition is real, and the achievement is real. The difficulty is for language to express one of them without explaining away the other.

—Alfred North Whitehead (1966, 102)

There is some sense—easier to feel than to state—in which time is an unimportant and superficial characteristic of reality. Past and future must be acowledged to be as real as the present, and a certain emancipation from slavery to time is essential to philosophic thought.

—Bertrand Russell (1917, 21)

Gather ye rosebuds while ye may,
Old time is still a-flying,
And that same flower that smiles to-day
To-morrow will be dying.
    —Robert Herrick, *To the Virgins to Make Much of Time*

Our destiny . . . is not frightful because it is unreal; it is frightful because it is irreversible and ironbound. Time is the substance of which I am made. Time is a river which sweeps me along, but I am the river; it is a tiger which mangles me, but I am the tiger; it is a fire which consumes me, but I am the fire. The world, unfortunately, is real; I, unfortunately, am Borges.

    —Jorge Luis Borges (1967, 64)

For us believing physicists, the distinction between past, present and future is only an illusion, even if a stubborn one.

    Albert Einstein (Cited in Hoffman, 1972, 258)

The ghost of time cannot permanently be laid.
    —Mary F. Cleugh (1937, 16)

Time present and time past
Are both perhaps present in time future,
And time future contained in time past.
If all time is eternally present
All time is unredeemable.
    —T. S. Eliot, *Burnt Norton*

It is a poor memory that remembers only backwards.
    —White Queen to Alice

It is a mere accident that we have no memory of the future.
    —Bertrand Russell (1921, 234)

In space-time, everything which for each of us constitutes the past, the present, and the future is given in block, and the entire collection of events, successive for us, which form the existence of a material particle is represented by a line, the world-line of the particle. Each observer, as his time passes, discovers, so to speak, new slices of space-time which appear to him as successive aspects of the material world, though in reality the ensemble of events constituting space-time exist prior to his knowledge of them.

—Louis de Broglie (1949, 113)

The *flow* of time is clearly an inappropriate concept for the description of the physical world that has no past, present and future. It just is.

—Thomas Gold (1977, 100)

Time is invention or it is nothing at all. But of time-invention physics can take no account. . . . *Modern physics . . . rests altogether on a substitution of time-length for time-invention.*

—Henri Bergson (1911, 361)

Time heals all wounds.

—Traditional saying

Time wounds all heels.

—Nontraditional saying

To realize the unimportance of time is the gate of wisdom.

—Bertrand Russell (1921, 226)

Apart from time there is no meaning for purpose, hope, fear, energy. If there be no historic process, then everything is what it is, namely, a mere fact. Life and motion are lost.

—Alfred North Whitehead (1966, 101)

[Time's] Moving Finger writes; and, having writ,
Moves on: nor all your Piety nor Wit
Shall lure it back to cancel half a line,
Nor all you tears wash out a Word of it.

—Omar Khayyám, *The Rubáiyát*

There can no longer be any objective and essential . . . division
of space-time between "events which have already occurred"
and "events which have not yet occurred." . . . Relativity is
a theory in which everything is "written" and where change
is only relative to the perceptual mode of living beings.

—Costa de Beauregard (1966, 429)

The tendency toward assimilation of time and space—which
is really . . . a transformation of time into space . . . exceeds
the authority of the most clearly established facts and the most
basic foundations of science.

—Emile Meyerson (1976, 356)

The relativistic union of space with time is far more appropriately characterized as a dynamization of space rather than
a spatialization of time.

—Milič Čapek (1976, 515)

Relativity physics has shifted the moving present out from
the superstructure of the universe, into the minds of human
beings, where it belongs.

—P. C. W. Davies (1976, 2)

The theory of relativity conceives of events as simply being.

—Adolf Grünbaum (1963, 388)

It is not the theory of relativity which "conceives of events
as simply being . . . ," as Grünbaum claims, but rather Grün-

baum, the philosopher, who conceives of the world as being thus.

—Richard M. Gale (1968, 218)

The objective world simply is. It does not happen. Only to the gaze of my consciousness crawling upward along the lifeline of my body does a section of this world come to life as a fleeting image.

—Herman Weyl (1949, 166)

Science in its effort to become "rational" tends more and more to suppress variation in time.

—Emile Meyerson (1930, 238)

It seems to me that it would be either a miracle or an unbelievable coincidence if all the major scientific theories . . . somehow managed to co-operate with each other so as to conceal time's arrow from us. There would be neither a miracle nor an unbelievable coincidence in the concealment of time's arrow from us only if there were nothing to conceal—that is, if time had no arrow.

—Henry Mehlberg (1980, 207)

Science can find no individual enjoyment in nature: Science can find no aim in nature: Science can find no creativity in nature; it finds mere rules of succession. These negations are true of natural science. They are inherent in its methodology. The reason for this blindness of physical science lies in the fact that such science only deals with half the evidence provided by human experience.

—Alfred North Whitehead (1966, 154)

Time sucks!
—Graffito (perhaps written by a student of a professor who maintains that time originates in black holes)

There is no other way to solve the problem of time than the way through physics. . . . If time is objective the physicist must have discovered the fact. If there is Becoming, the physicist must know it. . . . If there is a solution to the philosophical problem of time, it is written down in the equations of mathematical physics.

—Hans Reichenbach (1956, 16)

It cannot be too often emphasized that physics is concerned with the measurement of time, rather than with the essentially metaphysical question as to its nature. . . . We must not believe that physical theories can ultimately solve the metaphysical problems that time raises.

—Mary F. Cleugh (1937, 51)

Time is indefinable . . . due to the fact that temporal notions are implicitly involved in all of the basic concepts by means of which we think and talk about the world.

—Richard M. Gale (1968, 5)

It seems to me that in any operational view of the meaning of natural concepts the notion of time must be used as a primitive concept, which cannot be analysed but must be accepted, so that it is meaningless to speak of a reversal of the direction of time.

—Percy Bridgman (1955, 251)

The basic objection to attempts to deduce the unidirectional nature of time from concepts such as entropy is that they are attempts to reduce a more fundamental concept to a less fundamental one

—G. J. Whitrow (1980, 338)

What, then, is time?—if nobody asks me, I know; but if I try
to explain it to one who asks me, I do not know.
—Saint Augustine, *Confessions* (11, 14)

It is impossible to meditate on time and the mystery of the
creative process of nature without an overwhelming emotion
at the limitations of human intelligence.
—Alfred North Whitehead (1964, 73)

1. David Ray Griffin

INTRODUCTION: TIME AND THE FALLACY OF MISPLACED
CONCRETENESS

1.1. Time, Physics, and Metaphysics

In the title of this book, and especially in this introductory essay,
"time" means time as known and understood in human experience.
What exactly this is, of course, is by no means obvious! But there is
considerable consensus, even among writers who disagree radically
about the ultimate significance of time so understood, that time as
experienced involves at least the following characteristics: (1) a one-
way direction that is in principle irreversible, (2) categorical differences
between past, present, and future, and (3) constant becoming.

It is commonly held that physics provides no basis for any of these
features. The so-called fundamental laws of physics are often said to
require "time" only in the very abstract sense of the t-coordinate, on
which events are strung out. Not only do they provide no basis for a
distinction between past, present, and future, and a sense of an irre-
vocable ongoing; they do not (with possibly one minor exception)[1] even
allow for "anisotropy,"[2] the phenomenon in which the order of events
when read off in one direction can be objectively distinguished from
the order when read off in the other direction. And it is often considered
entirely a matter of convention that the vector points from the "past"
to the "future" (Davies, 1976, 12f.).

Thermodynamics does provide for anisotropy and hence, as K. G.
Denbigh (1981, 167) says, brings "the concept of 'time' a little closer
to what we have from our own experience." The thermodynamic
"gradient of monotonically changing entropy states . . . appears to
correlate completely with our own judgement of the temporal order of
events." However, according to the dominant interpretation (at least
prior to Prigogine's influence), there is still no distinction of past,
present, and future and no "one-way" going. "For although thermodynamics finds the two directions of time to be distinguishable, it does
not display the one direction as being in any sense 'more real' than
the reverse direction. . . . The question 'Which direction along the t-
coordinate is the real direction?' just doesn't arise in physical science"
(Denbigh 1981, 167). Furthermore, the anisotropy provided by thermodynamics is said to be a property of large numbers of entities or

events, not of these individual entities or events themselves, and hence it is considered a statistical matter.[3] Even at this level there is said to be no irreversibility *in principle*.[4]

In time as experienced, however, that today comes after yesterday, that earlier events come before later ones and not after them, is not a matter of interpretation of statistical data. Rather, the irreversibility of time seems to be analytic, a necessary aspect of what we *mean* by time.[5] There seems to something qualitatively different about time as we experience it and "time" as defined by theoretical physics or even thermodynamics,[6] according to heretofore dominant interpretations.

In time as we experience it, the past, present, and future are categorically distinct. The *past* is composed of all those events which are totally determinate, in which all options have been eliminated. As the epigraph says, once time's moving finger has writ, neither piety nor wit can cancel a word of it. The *future,* on the other hand, is that realm in which at least some indetermination remains; decisions remain to be made from among various possibilities. Although absolute determinists *say* that there are in reality no alternative possibilities, beyond what will actually occur, everyone *in practice* lives as if the future is in some sense open, as if there are real decisions to be made—which brings us to the present. The *present* is the realm in which decisions are being made: some possibilities are being turned into actualities while other possibilities are being excluded from actualization. So, we have these categorical differences: the past is the fully determined, the present is what is becoming determined, and the future is the still-to-be-determined. In classical physics there was no basis for these categorical distinctions. Some now say that quantum physics provides a basis for this tripartite division,[7] but this issue is still debated. The dominant view has been that fundamental physics is timeless. In any case, few thinkers deny that human experience of time includes this threefold division into past, present, and future—or at least the illusion of it.[8]

The term *constant becoming* is used to point to the fact that the present, or "now," does not "stand still," as some put it. More precisely, the present "now" never divides the same sets of events into past and future. In each new "now" there are events in the past that were not there before and that previously had been at most anticipated as possible, or perhaps probable, events. This is the feature of time that has been asserted by many writers to be most totally absent from physics. Physical theory is usually said to be indifferent to any idea of becoming, of events previously "in the future" coming into present existence.[9] This would be all the more the case if this idea were taken to include the idea that this coming into existence involves the transformation of potentialities into actualities.

These three widely agreed-upon aspects of time as we experience it,[10] at least as I have explicated them, all presuppose each other. Hence, none of them really provides any "stronger" sense of time than the others. For example, part of the reason that time is irreversible is that the past is composed entirely of totally settled, determinate events; it would be self-contradictory for those events somehow to come to be in the future, for that would mean they would be in the realm of the still partially indeterminate. Likewise, it is because the present involves the transformation of possibilities into actualities that there is constant becoming, which means that time cannot "stand still." Once some possibilities have been chosen and others excluded, one cannot face these same options. Those decisions are, as we say, now "behind us," and we now confront new options.

This book is concerned with the ultimate significance of time in this sense. The epigraphs at the head of this chapter were chosen to illustrate several points. One is that humanity has harbored greatly different attitudes about time, with its irreversibility and incessant becoming. Some people focus on the aspect of time that leads them to call it a "devouring tyrant"; others see it as a "refreshing river." Some see it in theodical terms, as ultimately healing all wounds and wounding all heels; others see its "perpetual perishing" as *constituting* the ultimate problem of evil. Some passionately believe the universe would be an aesthetic failure unless it were symmetrical in all respects; others with equal passion believe that unless the relations of the present to the past and to the future are asymmetrical, so that we have genuine freedom partially to create the present and hence the future, life would be meaningless. Some believe that a life of wisdom and happiness requires coming to terms with the ultimate significance of time; others see it as requiring an appreciation of time's insignificance, even illusoriness. These various attitudes toward time are not directly the subject of this volume. However, they are not irrelevant to the subject, since what we *want* to believe, for moral, religious, or aesthetic reasons, usually conditions the types of evidence to which we attend and the relative weight we give to the various considerations.

A second purpose of the epigraphs is to illustrate the point that thinkers differ greatly on the subject of this volume: the ultimate significance of time. Finally, many of the epigraphs directly reflect the more complete topic of this volume: the bearing of physics upon the question of the ultimate reality of time. Here again, thinkers differ drastically. They can be divided into three major groups.

First, there are those who hold that time is to be ascribed no other nature, and no more reality, than it has in the fundamental laws of physics. This position has had much strength in the last two centuries, because of widespread acceptance of the metaphysical position of reductionistic materialism among physicists and philosophers (even if it was not acknowledged as a "metaphysical" position) and because of widespread abdication on the part of philosophers from involvement in the philosophical issues in physical theory. The majority of philosophers who *have* concerned themselves with these issues, the "philosophers of science," have done so from a viewpoint (e.g., operationalism) that has sanctioned the tendency of physicists to give terms no larger meaning than that vouchsafed by the methods, data, and theoretical formulations of their own discipline. In any case, those holding this position have tended to speak of time as an illusion (see the epigraphs by Russell, Einstein, de Broglie, Gold, de Beauregard, Grünbaum, Weyl, Mehlberg, and Davies).[11]

A second position is constituted by those who agree with those in the first group that time, like all meaningful concepts, must be defined by science, but who hold that, although the concept of time is not grounded in the fundamental laws of physics, it *is* grounded in some other feature of the world that is susceptible to scientific treatment. The most popular form of this position is the view of time as rooted in the laws of thermodynamics. From this point of view, it is arbitrary to say that time is unreal simply because it is not reflected in the "fundamental (timeless) laws" of the universe. Science also deals with historical contingencies and so can include the "initial conditions" of the universe (which are presumed to be the source of "time's arrow") among its fundamental data. From this viewpoint, it is fundamentally the direction of entropy that defines the reality of time. Our own experience of time is taken to be derivative from the fact that we are composed through and through of entities involved in the universal entropic process.[12]

However, other approaches are possible within this general position. Time could be regarded as first arising in living processes and hence would be established by the biological sciences. Or it could be thought a mind-dependent property and hence to have arisen only when organisms with minds sufficiently analogous to our own had emerged (often thought to occur only with the development of a central nervous system). Proponents of these views would tend to stress the reality of time to the degree that they consider living, or psychological, processes as genuinely real—not some frothy epiphenomena on the surface of the really real, strictly physical, processes.

In fact, within this second position as a whole the sense in which time is spoken of as real is a matter of degree, seeming to depend not only upon the author's systematic position but also upon his or her desire to regard the universe as essentially temporal or timeless. None of those in this group regard time as of strictly ontological or metaphysical significance, i.e., as rooted in the very nature of reality. They all see it as something contingent, as something that depends upon particular features of our universe that conceivably could have failed to occur—or at least they believe that they *as scientists* can only so regard time. Since time is therefore regarded as a contingent feature of reality, it can be regarded as more or less "real" or "illusory," depending upon the other interests and biases of the author. Hence, the difference between this and the first position is ultimately a difference of degree, so the question of which writers would be classed in the second as opposed to the first position is somewhat arbitrary. However, it is a difference of degree that becomes virtually a difference in kind, as some advocates of this second position stress the reality, universality, and significance of time so strongly as to make their difference from the third position discernible only by careful reading. This would be true, for example, of Ilya Prigogine—if he is to be understood as exemplifying the second rather than the third position (see section 1.3).

The third position is constituted by those who hold that it is not the task of one of the special sciences to define and account for the reality of time. For one thing, the methods adopted by the sciences are, at least thus far, said to be such as to preclude treatment of the temporal features of reality; hence, looking to physics to learn about time is like asking a computer to answer a question it has been programmed to ignore. Furthermore, every attempt to explain temporality in terms of some physical characteristic of the world involves, it is claimed, the fallacious procedure of reducing a more fundamental concept to a less fundamental one (see epigraphs by Gale, Bridgman, and Whitrow). Within this third group, some would say that time is simply a "primitive" concept that need not and probably cannot be explained in any sense, either physically or metaphysically. It suffices to see that time (as, e.g., Gale would claim) is an inevitable feature of our experience, as reflected in our ordinary language, which is presupposed by all our technical languages, so it is senseless to try to explain it away on the basis of any of those technical languages. However, some within this third group would say that time is a topic for metaphysics, or ontology,[13] meaning thereby that an approach is required that does not limit itself to the methods and abstractions of any of the special sciences, but that attempts to synthesize the presuppositions and results of all the special sciences with each other and with the knowledge and especially presuppositions of human experience in its fullness (which may include features not in the province of any of the special sciences).

1.2. Time in Process Philosophy

There are many ontologies or metaphysical systems that fit the general characterization given in the previous sentence. (There are also some that do not, since they ignore the results of the special sciences; but they are themselves ignored here, just for that reason.) This introductory essay is written from the point of view of one of them, the "process philosophy" derived primarily from the writings of Alfred North Whitehead. From this perspective, the natural sciences, at least as practiced thus far, methodologically abstract from the full concreteness of the entities or processes they study. Therefore, to jump from the mere fact that time is not present in natural science to the conclusion that time is not real at the fundamental level of nature is to commit the "fallacy of misplaced concreteness." The fallacy is to treat the abstractions from certain things—abstractions focused on because of certain interests and methods—as if they were the concrete things themselves.[14] It is to treat the map as if it were the territory, assuming that what is not on the map is not in the actual terrain itself.

The thesis implied by the title of Whitehead's major book, *Process and Reality,* is that process *is* the concrete reality of things and, conversely, that concrete realities *are* processes.[15] From this viewpoint, time, or temporality, is an ultimate feature of reality. It is not itself an actual or concrete entity; it is a *relation*—a relation of conformity to and inclusion of the past. This is a version of the relational theory of time. However, many relational theories make time a contingent feature of reality, since time is presumed to depend upon a kind of relation that may or may not exist. For example, time in Augustinian theology—according to which the world was created *ex nihilo* in a sense that made the existence of finite things as such purely contingent, dependent upon the voluntary decision of God—was purely derivative, a relation among such contingent things. Of course, if the everlastingly existent creator had itself been a temporal being, as in Isaac Newton's thought (as Milič Čapek points out),[16] then time could be considered to be a relation and yet everlastingly real, as it would most fundamentally be the relation between the creator's successive states of consciousness. (Most commentators have ignored Newton's heterodox theology, and his talk of "absolute time" has generally been misunderstood to mean that time is not in any sense a relation and hence can exist apart from actual events.) But since traditional theology held God to be nontemporal in every sense, time was itself considered part of the contingent creation. Much modern science-based philosophizing, in thinking of time as a derivative, emergent reality, has thereby held to this fundamental dictum of traditional theology, that time or process has no ultimate reality.

But process philosophy, while it speaks of the present world as creation, emphatically rejects this notion of *creatio ex nihilo*, according to which there might be no plurality of things whatsoever, and the related idea that temporality or process is a derivative feature of the world. Although time is a relation, there have always been processes or events having the twofold structure of inclusion and anticipation (i.e., anticipation of being included by future events). The existence of such events or processes is hypothesized to be ontologically necessary, simply part of the ultimate nature of things, which just *is*.[17] In answer to the question, for example, What existed before the "big bang?" process philosophy would claim—assuming the correctness of this theory of the origin of our universe—that such events or processes existed. Hence, process philosophy can be characterized as pantemporalism.

From this perspective, the relations and objections of process philosophy to the other basic options for the ultimate reality of time can be clarified. The three options would be: reductionistic nontemporalism, dualism, and pantemporalism. Reductionistic nontemporalism can be summarized in the following three propositions. (1) The fundamental laws of physics tell us all the fundamental characteristics of the universe. (2) These laws do not include time. (3) Therefore time is unreal, an illusion fabricated and projected onto reality by human experience.[18]

One objection of process thought to this reductionistic position has already been suggested: there is no good reason to suppose that the fundamental laws of physics, especially of physics in its infancy (a few hundred years is *not* very long), give us all the fundamental characteristics of the universe, even of the most "elementary"[19] entities or processes constituting it. There is much more reason to suppose that physics, at least as developed thus far, involves significant abstractions from the concrete realities of things. A second objection is that, according to the reductionistic viewpoint, human experience, with its inherently temporal structure, is an epiphenomenon, not a full-fledged actuality; and yet it is credited with the power of creating the illusion of time. This attribution of such enormous creative power to unreal things should be taken as a *reductio ad absurdum* of reductionism. A third objection is based on an idea that process philosophy shares with the "commonsense" and "pragmatist" schools of philosophy. This is the idea that the ultimate criteria for testing philosophical doctrines are those notions that all people in fact presuppose in practice, even if they deny them verbally. I call these the "hard-core commonsense" notions, to distinguish them from those ideas which are often thought of as "common sense" but which belong to the "soft core" of one's repertoire of common sense (that is, they are really rather provincial ideas and are definitely not ones that *all* human beings *inevitably* presuppose in all they do). Because the hard-core commonsense notions cannot be consistently and hence meaningfully denied—since the very attempt to refute them would presuppose them—they should provide the ultimate constraint upon philosophical opinions.[20] We should accept no doctrine in our theory that we cannot consistently live by in practice (as the pragmatists put it).

The doctrine of the ultimate reality of time is one of those doctrines.[21] The very act of living, including the activity of articulating and defending scientific and philosophical ideas, necessarily presupposes the dimensions of time as described above: that we can affect the future, but not the past; that there is an irreversible directionality to reality; that the present involves turning some possibilities into actualities and excluding others; and that—however inadequately this formulation puts it—the present "now" constantly moves. Rejecting this temporal aspect of reality as an illusion is done on the basis of accepting the truth of some other proposition as the major premise for a syllogistic argument. For example, the proposition used above was: "The fundamental laws of physics tell us all the fundamental characteristics of the universe." This is obviously *not* one of the "common notions" that all human beings inevitably presuppose in practice. On what grounds do we say that we are so certain that this proposition is true that we will deduce from it the falsity of a proposition that is among the universal presuppositions of human practice?[22]

There is a fourth basis on which process philosophy considers reductionistic nontemporalism unworthy of belief: the widespread acceptance of a materialistic metaphysics and of a corresponding sensationalistic epistemology, according to which anything knowable is knowable through the physical senses, has led to the supposition that all meaningful concepts apply only to objects of sensory experience. But our ordinary notion of time is emphatically *not* derived from sensory experience; we do not see (hear, smell, taste, or touch) the past, we do not see unactualized possibilities, we do not see a moving "now." Since from a sensationalist perspective, as Richard Gale points out (1968, 69f.), the *factual* content of all assertions can be equated with their *sensible* content, it is assumed that assertions about time can be stripped of those features arising from nonsensory experience without loss of anything essential. However, Whitehead (1966, 72–75, 133, 152–62) argues that scientific theory presupposes a set of concepts, time being only one among them ("causation" is another), that is rooted in a kind of perception (which he calls "prehension," or "perception in the mode of causal efficacy") that is not sensory and that is in fact more basic than sensory perception. Sensory perception provides *precision* of observation, but not the *meaning* of the concepts employed in scientific theory. Once it is realized that scientific theory in general cannot survive on the basis of a rigidly applied sensationalist epistemology, there should be less compulsion to regard as "phoney"[23] those characteristics of time that are derived from nonsensory aspects of experience, such as memory and anticipation.

The second position in contrast to pantemporalism, dualism, might at first glance seem the most reasonable. According to this view, there is no time, no temporal process, at the level of the most elementary entities constituting nature, such as that of individual atoms and subatomic particles. Time is said to emerge only as a result of the emergence of certain conditions or types of events. Some dualists, as indicated earlier, locate this at fairly high levels, e.g., with the emergence of life, consciousness, or, in extreme cases, *human* consciousness. But others locate the emergence of time at a much more primitive level, in some cases so primitive that to insist upon a distinction between this view and that of pantemporalism may seem like quibbling. However, there *is* an important philosophic problem involved—the same problem that plagues every form of ontological dualism. This is the problem of how two *ontologically* different types of entities can *interact*. The best-known form of ontological dualism is the one responsible for the "mind-body" problem, the Cartesian dualism between experiencing mind and non-experiencing matter. Most philosophers have given up this view of the mind-body relation precisely because it has proved impossible to understand how mind and body, so understood, could interact (at least without resorting to the appeal, less acceptable now than in the seventeenth and eighteenth centuries, to a divine omnipotence that is not deterred by mere impossibilities). But the problem of understanding how temporal and nontemporal things could causally interact (*emergence* presupposes a type of causal influence) is at least as difficult, as Capek points out.[24] How can we even think coherently about events for which the distinction between past, present, and future obtains as interacting with events (or things, since the term *events* already presupposes time) for which this distinction does not apply? This difficulty is one of the major reasons process philosophers urge pantemporalism as a more reasonable view.[25]

Process philosophy's pantemporalism is articulated in several essays in this volume, including those by Bjelland, Hurley, Stapp, Cobb, Čapek, and me. However, the crucial point, mentioned earlier, needs to be developed here. For Whitehead, the reality of time, with its irreversibility, is based on the fact that *the actual world is composed exhaustively of momentary events that include, partially but really, preceding events,* which had in turn included previous events, and so on back. In Whitehead's words: "This passage of the cause into the effect is the cumulative character of time. The irreversibility of time depends on this character" (1978, 237).

Because of this character of events, a present event is *not* independent of previous events; rather, it presupposes just those events, since it includes them and is largely constituted by this inclusion. Because of this essential inclusion of prior events, the idea of successive events occurring independently of each other arises only by abstracting from the full reality of the events.

Time in the concrete is the conformation of state to state, the later to the earlier; . . . pure succession is an abstraction of the second order, a generic abstraction omitting the temporal character of time. . . . The immediate present has to conform to what the past is for it, and the mere lapse of time is an abstraction from the more concrete relatedness of "conformation." (Whitehead, 1959, 36)

It is therefore a fallacy to think of the real events or things making up the world as having the property of "simple location," which would mean that they could be satisfactorily described without reference to prior and following events.[26] This is no small point, since the traditions of Humean and thereby Kantian philosophy have presupposed just this idea of events as "simply occurring."[27] That is, the events in themselves were held to have no inherent conformal relationship to prior events. These philosophical traditions are thereby based on the fallacy of misplaced concreteness.

The same is true of those materialisms or dualisms that think of nature as composed of "self-contained particle[s] of matter."[28] Rather than thinking of enduring particles as the fundamental entities of the world, Whitehead sees each enduring object as composed of a rapidly occurring series of events, each of which includes aspects from its predecessors in that enduring object but also aspects from other prior events as well. Hence, Whitehead opposes the widespread view that an individual atom is timeless; rather, it is a "temporally-ordered society" of actual occasions.[29] Again, the mistake is to confuse an abstraction (the *form* of energy, which remains relatively the same through time) with the concrete reality, which is a series of momentary events, each of which derives its form largely through conformity with predecessors and passes it on to successors. Since an individual atom (or even electron) has a temporal structure, time or temporality does not first arise as a statistical effect of the interactions among a multiplicity of atoms.

I previously drew an analogy between three views on time and three views on the mind-body problem. I now suggest that the relation between them is more than an analogy. The two issues are finally one. I will introduce this idea by reference to the thought of a philosopher who argues for the ultimate unreality of time in the physical world. Adolf Grünbaum contends tirelessly that time, in the sense of becoming, is a mind-dependent property. From this he concludes that it cannot exist in the physical universe. However, although he sometimes speaks as if he meant only human minds, since he speaks of time in the sense of becoming as "anthropocentric," in more careful formulations he makes clear that he does not limit the requisite mind to that of human beings (Grünbaum, 1967, 152, 179–80). Although most process thinkers see Grünbaum as one of their staunchest opponents,[30] as indeed he is in most respects, I want to stress the agreement between his thought and process thought on the mind-dependent nature of time. Of course, process philosophers generalize "mind" much more broadly than Grünbaum would countenance; he is hesitant even about cockroaches (Grünbaum, 1967, 179f.), whereas process philosophers generalize "mind-like"—or, better, "experience"—to all actualities whatsoever. We agree with Grünbaum that time is experience-dependent. But since we hold that *all* actualities are units of experience and that a plurality of such actualities necessarily exists, we hold that time exists necessarily, not as a contingent emergent. ("Actualities" here refers to genuine individuals, not aggregates; see 9.3.2, below.)

Accordingly, process philosophy's pantemporalism and its panexperientialism are ultimately one and the same thing (as Andrew Bjelland's essay makes clear). Likewise, it is consistent for reductionistic materialism, which denies experience except as an epiphenomenal illusion, to deny time except as an illusion created by an illusion. And it is consistent for dualists on the mind-matter issue to be dualists in regard to time.[30a] Finally, process philosophy's pantemporalist, panexperientialist position stands as a clear alternative both to reductionistic, materialistic nontemporalism, and to temporal-nontemporal, mind-matter dualism. It holds that temporality as such and experience as such do not emerge at some point, that becoming is not somehow generated from nontemporal being, nor experience from nonexperiencing matter; indeed, it holds that such generations are unthinkable. However, in contrast with reductionism, it does not thereby regard experience and time as illusions that magically appear at some point. Accordingly, it denies that it is the task of a special science, whether physics or some other, to determine when and how time, or its illusion, first appeared.

The relevance of the fallacy of misplaced concreteness to this discussion of the relation between panexperientialism and pantemporalism must now be clarified. Whitehead, in one of the quotations at the head of this essay, speaks of that science which finds no creativity (and hence no time) in nature as ignoring half the evidence. He means this literally, for he regards each momentary event as having two modes of existence. It comes into existence as a process of becoming (called "concrescence," the process of becoming concrete). In this mode of being it has experience—not usually self-consciousness, or even consciousness, but an emotional internalization of the environment nevertheless. In this mode it is a subject. This is what the event is in and for itself. In this mode, it is known only to itself; it cannot, in principle, be perceived by any other subject. However, this mode of existence is quickly over—it may last from (perhaps) a tenth of a second in some cases (e.g., in human experience) to a billionth of a second in others (e.g., in a subatomic particle that vibrates a half-billion times per second).[31] When this process of concrescence is completed, the event begins its second mode of existence: it becomes an object for others, an object to be felt by subsequent events—which might include being perceived by a human observer or the instruments thereof. In other words, it becomes a cause influencing later events.

Now, when Whitehead says that it is inherent in scientific methodology to ignore half the evidence, he means that such methodology considers things only as they are perceivable—either directly or by means of magnifying instruments. This means that it considers things only when they are objects, *after* their period of subjectivity, or experience, is over. In other words, it considers things only on their extrinsic side and ignores their intrinsic natures. It may consider a "thing" as a process, but it deals only with its *extrinsic* process, which Whitehead calls "transition," since there is a transition of energy from the one event to another. But it ignores the event's *intrinsic* process, which is "concrescence." In fact, as Whitehead states elsewhere, physics does not even deal with half the evidence, since it contains the extrinsic side of things only in regard to their spatiotemporal effects upon other things (1926, 220). In any case, all the features of time discussed earlier are rooted in the *intrinsic* reality of events, in the process by which they become concrete, or determinate, for it is here that the event includes the past events into itself and it is this inclusion that makes time irreversible.[32] Accordingly, any approach that commits the fallacy of misplaced concreteness by equating the extrinsic side of events with their complete reality will necessarily miss the roots of time in those events.

Behaviorism was the result of the decision that, to be a science, psychology had to ignore the intrinsic reality of human beings, which is known only by being one, in favor of looking exclusively at the features of human beings that can be known from without, through the sensory perception and magnifying instruments of the exterior observer. Those who promoted this form of psychology tried to get along without using *experience, consciousness, purpose, aim, desire, feeling,* or any other "subjective" terms. They wanted to exclude all "anthropomorphic" categories from the study of human beings, so that psychology could finally become a genuine science, like physics and chemistry.[33] But now it is generally accepted that an exclusively behavioristic approach cannot deal with the most important features of human beings. If we think of human experience as fully natural, why should it be excluded from psychology as a natural science? The most adequate approach to the study of human beings comes from combining behavioral and introspectionist methods.

But if this point is accepted, and we are nondualists, should not the twofold approach be extended to all levels of actuality? Of course, we cannot ask nonlinguistic entities what it feels like to be one of them in order to get at their intrinsic natures. But likewise we need not study them on the basis of the assumption that they *have no* intrinsic side, that they *have no* experience, that they are nothing but objects. For that assumption is pure speculation, no less "metaphysical" than its contrary. By reasoning from analogy—the grounds of which have been greatly strengthened by acceptance of an evolutionary perspective—it is more reasonable to speculate that all individuals have an intrinsic nature, with some degree of experience, than to speculate that some of them do not, for *we* do; we know this directly. I know that there is more to me than the behaviorist psychologist can describe. And you know the same about yourself. And you presume it about me in practice, no matter how solipsistically you may rhapsodize in theory. Is it not less arbitrary to assume that this twofold mode of existence applies to *all* individuals, rather than to assume that it "emerged" at some point in the evolutionary process? To assume the latter would be to assume an ontological dualism—no matter how vehemently one might reject the label. To say that some things have an intrinsic as well as an extrinsic reality, and that other things have only an extrinsic reality—i.e., to say that some things are subjects as well as objects, whereas other things are only objects—is to affirm an ontological dualism, regardless of the name (such as "mind-brain identism") with which one seeks to disguise it.[34]

The widespread idea that physical events could in principle start running backwards presupposes the idea that temporal order is pure succession. We know that it is nonsense to think of our own experience as running backwards because we know, at some level, that the order of our experiences is not that of pure succession, but of the partial *derivation* of later from earlier experience, with partial *conformation* of later to earlier. In Whitehead's words:

Time is known to us as the succession of our acts of experience. . . . But this succession is not pure succession: it is the derivation of state from state, with the later state exhibiting conformity to the antecendent. Time in the concrete is the conformation of state to state, the later to the earlier; and the pure succession is an abstraction from the irreversible relationship of settled past to derivative present. (1959, 35)

Most people would agree that this correctly describes human experience, but they do not see how it could apply to "purely physical" processes. Yet Whitehead suggests that it does: the deleted words in the above quotation are: "and thence derivately as the succession of events objectively perceived in those acts." Because he, on the basis of an evolutionary, nondualistic outlook, takes unitary "physical events" to be not different in kind from the "mental events" constituting our own immediately known experience, he can hold that time as "known to us in the succession of our acts of experience" can be attributed *by analogy* to "the succession of events objectively perceived in those acts." I put "physical events" and "mental events" in scare quotes to indicate that this dualistic language is mistaken from the perspective of process philosophy. The difference between the proton and the psyche is one of degree, not of kind (in an ontological sense). One who holds otherwise is a dualist, no matter how odious such a designation may be.

If this outlook were to become the accepted framework (or paradigm) within which science is practiced, in place of the ontological materialism and dualism between which most scientific work in the modern period has oscillated, one of two consequences would follow for the question of the reality of time in physics. One possibility would be that physicists would retain the methodological limitation to the extrinsic nature of events. Only now they would see this to *be* a self-imposed limitation and would assume that, apart from the aspect of events upon which they focus *qua* physicists, the events also have an intrinsic side, which is analogous in some remote way to our own experience and in which the basis for temporal relations at the most elementary level of nature is located. Hence, they would be freed from the assumption that it is their task, *qua* physicists, to define the meaning of time and account for its existence—or, failing this, to deny its reality.

A second possibility would be that they would reconceive the nature of physics so that it could, in principle, take account of the intrinsic as well as the extrinsic nature of events. If this occurred, the result could properly be called the beginning of "post-modern" science, since it can well be claimed that the most significant feature distinguishing modern from pre-modern science is the exclusion of all categories of subjectivity or experience from consideration, either as data or for use in explanatory theory.[35] However, the resulting science would be *post-modern*, not a return to pre-modern, insofar as the gains in rigor and precision acquired during the modern period were retained.[36] In such a science the reality of time, with its features as known through human experience, could be included within science itself.

1.3. Time in the Work of Bohm and Prigogine

I turn now to the question of how the programs of Bohm and Prigogine are related to the issues involved in physics and time as seen from the perspective of process philosophy. Bohm seems closer to developing a post-modern physics in the above sense. He, like Whitehead, understands apparently enduring particles to be "world tubes" composed of rapidly occurring series of events. Furthermore, he sees each momentary event as enfolding or "implicating" the whole of reality within itself. Finally, he says that physics thus far has stressed the "explicate order," but that it now needs to deal also with the "implicate order." At least in some of his formulations, this implicate order seems to involve that phase of the momentary events in which they enfold the whole of reality within themselves.

Thus, Bohm would seem to be developing a physical theory corresponding to Whitehead's metaphysical hypothesis that each momentary event has a subjective side and that this subjective side is a process of concrescence in which the whole past world, under abstraction, is included. However, sometimes Bohm has spoken as if the "whole" that is enfolded in each event were not simply and directly the past events, but a nontemporal order in which the future as well as the past is contained. If this were his position, the reality of time would be denied, for from the ultimate point of view (i.e., of the "super-implicate order"), what is future to us would already (eternally) be as determinate as what is past. I will not discuss this and other issues in Bohm's thought any further here, however, since they are explored in my essay and Cobb's, below.

Prigogine's position presents the process philosopher with something of the inverse situation. On the one hand Prigogine rejects, at least so far, any revision of the basic nature of natural science so as to allow it to include subjectivity among its data or to speak of any aspect of events that is not in principle observable. On the other hand he is as intent as any process philosopher on insisting upon the ultimate significance of time, and this has led him to develop a "post-classical" science.

Prigogine indicates that the problem of time has been at the center of the research he has been pursuing all his life (Prigogine and Stengers, 1984, 10).[37] He says that the problem of the "two cultures" has not been that scientists have not read enough humanities and humanists enough science, but that there has been nothing in common between the two thought worlds. At the root of the cleavage has been the fact that, while the humanities and social sciences are necessarily time-oriented, classical science has been nontemporal (xxvii). Furthermore, this nontemporal view has been part and parcel of an alienating science that has portrayed a dead, debased nature, creating an inevitable opposition between humanity and nature (3, 4, 6, 89). This portrayal of nature has led, by reaction, to antiscientific metaphysics (3, 7, 31ff., 79). Overcoming the dualism between physics and philosophy, between matter and life, between science and the humanities, between nature and humanity, requires an enlarged science, with a new idea of time (14, 96, 175, 298). This in turn involves a new concept of matter, a return in a sense to one aspect of pre-modern naturalism, in which matter was seen as active and self-organizing (22, 32f., 36, 75, 82, 287, 291). Matter now should be seen as capable of "perception" and "communication" (33, 145, 163, 171, 180, 181) and as manifesting behavior that depends upon its special past history (153, 161). Furthermore, we should beware of speaking of "elementary" particles, since this term suggests an autonomy from context that is untrue (287).

All of this adds up to a post-classical science, taking "classical" science to be that which emphasized time-independent laws and deterministic processes, thereby portraying nature as a grand tautology

(2, 77, 213). The new science, which forges a new alliance between humanity and nature (to replace the animistic alliance broken by modern or classical science), stresses nondeterministic processes, in which there is intrinsic randomness—i.e., randomness due not only to our ignorance of deterministic causes (234, 239, 298)—for only genuinely nondeterministic processes can be irreversible (xxvii, 16, 276, 277). The fundamental character of irreversible processes is Prigogine's key thought; only this kind of process gives science the kind of time that is presupposed in all human experience, and hence in philosophy and the humanities in general. Prigogine's program is to reveal the existence of irreversibility at all levels, including and especially the microscopic level (16, 232, 258, 288, 289): there must be something at the microscopic level that provides the root of irreversibility at the macroscopic level. This latter irreversibility cannot intelligibly be thought to emerge, as a miracle, from fully reversible processes (285, 289, 298).

On all these points, Prigogine is in complete agreement with Whiteheadian process philosophy. Indeed, he often stresses his agreement with Whitehead as well as with Bergson (93–96, 129), seeing his own task to be that of giving scientific content and precision to their metaphysical speculations (24, 310). However, beyond these agreements there are some differences. These differences can be read as merely methodological—reflecting the different criteria and resources to which Prigogine, *qua* physicist, and Whitehead, *qua* philosopher, can and must appeal—or the differences could be read as more radical.

I will begin with the more radical interpretation. According to this reading, whereas Whitehead would exemplify the third position on the bearing of physics upon the ultimate reality of time (see above, section 1.1), Prigogine would be seen as exemplifying the second. The crucial difference would be whether human experience as such were used to establish the fundamental meaning and nature of time. For the later Whitehead (see Hurley's essay, below) it is, and the irreversibility in *nonhuman* nature is accounted for by the postulate that it is composed of processes that are *analogous* to our experiential process.[38] The adoption of this position by Whitehead was part of his move from the "philosophy of nature" in his pre-1925 works, in which the human subject is excluded and nature is regarded only as the object of sense experience, to "metaphysics," in which experience and the experienced are to be integrated into one scheme of thought, and inquiry about nonhuman events is extended beyond the description and correlation of their appearances to the question of what they might be in themselves.

Prigogine, on the other hand, sometimes speaks as if a post-classical science can reconceive time and nature adequately without breaking with classical science's dictum that "the soul which counts" lies outside the province of physical science (Prigogine and Stengers, 1984, 22). He sometimes (e.g., near the end of the paper in this volume) seems to account for *our* experience of time on the basis of the fact that we are examples of those highly unstable dynamic systems in which random-ness and hence irreversibility arise—which seems to make his procedure the opposite of Whitehead's. Consistent with this interpretation of Prigogine's position is the fact that he sometimes appears to regard time as a *contingent* rather than universal feature of the physical world because it requires a minimum complexity (16, 239, 251, 298, 301). There was, he says, a movement "from being to becoming"; i.e., becoming or time-irreversibility originated in the first stages of our universe (xxx, 278, 298, 300, 310). Hence, the forward direction of things is only a tendency, not a necessary truth (xxvii, 300). The world, accordingly, is pluralistic (xxvii, 251), or what I earlier labeled "dualistic," involving a mixture of reversible and irreversible processes (xxvii, 251, 232, 258, 289).

Under this interpretation, Prigogine would differ only in degree from previous thinkers who regarded the task of physical science to be to establish the *meaning* of time and who considered time a *contingent* feature that arose at some point in the development of our universe. Prigogine would, of course, differ *greatly* in degree, since he stresses the priority of irreversible processes, in contrast to those who have regarded irreversible processes as exceptional and hence have had to see the origin of life, with its irreversible processes, as a virtual miracle. Prigogine stresses that irreversibility occurs at every level of nature (285) and that reversibility is based on highly *artificial,* simplified situations: nature in the raw is not simple even at the most fundamental level, and in nature irreversibility is the rule, not the exception (8, 9, 10f., 215f.).

But even on this point (to continue this interpretation of Prigogine's meaning), the difference with Whitehead, while it may seem subtle, is important. For Whitehead, reversibility is the result not of the *artificial,* but of the *abstract.* That is, there is no true reversibility, not even in artificial, isolated conditions. The idea that time could be reversed results from mistaking abstractions for the fully concrete entities. In Prigogine's framework, however, it seems that genuine reversibility can occur: in artificially constructed contexts there can be processes that are fully deterministic, hence symmetrical and reversible. If time is *defined* in terms of the direction of physical processes, one would have to say that in those situations time itself is reversible.

The crucial question in regard to the correctness of the above interpretation of Prigogine's position is just this, i.e., whether he means that the far-from-equilibrium processes provide the basis for a *definition* of time. If he does, then, even though he says that the entropy barrier is infinite so that reversibility could never *in fact* occur, his position would suffer from the paradoxes afflicting all the other attempts to define time in terms of physical processes (see note 4, above).

However, there is another way to interpret Prigogine's position. This interpretation is based on his statement that the distinction between past and future is a primitive, i.e., prescientific, concept, which science must simply presuppose (1973, 590; 1980, 213). Under this interpretation, Prigogine's view would be not that it is science's task to provide an adequate concept of time, but that science must become consistent with the primitive "commonsense" notion of time we have from our own experience. Prigogine's work would move physical theory toward this consistency by showing that reversible time is not a good tool even for complex dynamical systems and that irreversibility is primary at every level of reality.

In harmony with this interpretation, the movement "from being to becoming" would not be a physical one (which would make irreversibility contingent), but a conceptual one, signifying the move from a physical theory in which simple, reversible systems have an autonomous and fundamental status within complicated systems to a theory in which they become singular limiting cases of an asymmetrical model. The pluralistic "mixture" of which Prigogine speaks would not refer to a true or *ontological* mixture, but would reflect his desire to make sense of the *phenomenological* difference that made classical dynamics possible, i.e., that some systems in fact behave—as a first approximation—in a reversible way. Finally, under this interpretation Prigogine's "artificial" would be the same as Whitehead's "abstract." That is, "artificial" systems would be those which had been *conceptually* simplified to make them conform to the conceptual tools of a reversible dynamics.

According to this second interpretation of Prigogine's meaning (which I owe to the coauthor of *Order Out of Chaos,* Isabelle Stengers),[39] there would be no difference in principle between the positions of Whitehead and Prigogine. They would agree that the basic meaning of time is rooted in a primitive level of experience, which it is the task of metaphysics to articulate. The differences would be due solely to the fact that, whereas Whitehead moved explicitly into a metaphysical or ontological context, Prigogine has sought to remain within the constraints of science, in which the meaning of all terms must be defined operationally (without confusing this operational meaning with an ontological one).

Whichever interpretation be the correct one, Prigogine's work is extremely significant from the viewpoint of process philosophy. For one thing, under either interpretation it goes a long way toward overcoming the dichotomy between the "two cultures," insofar as this dichotomy has been based on the assumption that physics, as the basic natural science, is essentially nontemporal. Also, Prigogine's work provides some verification of the Whiteheadian hypothesis, for if this hypothesis is correct—that cumulative and hence irreversible processes constitute the very nature of actuality as such, and hence of the processes at every level of our particular world—then the more physics advances, the more evidence of irreversible processes it should find. That is, the metaphysical or ontological irreversibility pointed to by Whitehead should give rise to precisely the kind of empirically detectable irreversibility focused upon by Prigogine.

If the second interpretation of Prigogine is accepted, then his position and Whitehead's are complementary, as physics and metaphysics, each providing what the other cannot. But if the former interpretation of Prigogine be accepted, then there is a basic philosophical difference: whether "time's arrow" is to be regarded as rooted in a kind of process that is irreversible not only contingently, and hence whether at the deepest level of the world irreversibility reigns supreme. This could be regarded as the issue as to whether the very meaning of time is to be provided by metaphysics, as Whitehead holds, or whether physics begins with a metaphysical notion of time but then has the task of improving upon it, as Prigogine sometimes seems to say (1973, 590).

However, there is another way of viewing the fundamental issue. Instead of a contrast between science and metaphysics, a contrast between a modern and a post-modern science could be entertained. From a post-modern perspective one could challenge the modern assumption that science cannot speak of subjectivity or experience, that the human soul and its analogues must be left outside of natural science as if they were, as in Descartes's dualism, outside of nature itself. Whitehead himself was ambivalent on this subject. He often spoke as if natural science had to remain with the categories of objectivity (see note 31, above); his protest against the fallacy of misplaced concreteness was a reminder that, to get at the fuller truth, we must include the scientific description of nature within a fuller metaphysical account (e.g., 1966, 18, 156). However, Whitehead also said that science's categories are not irreformable, and that he was raising his protest on behalf of science itself (1926, 121, 122, 128). In this latter mood he seemed to be saying that a more inclusive science could include categories of subjectivity, such as "experience," "concrescense," and "prehension." Prigogine's drive is certainly to find time rooted more and more deeply in the nature of reality. For example, he has said: "It would be quite appealing if the atoms' interaction with photons (or unstable elementary particles) already carried the arrow of time that expresses the global evolution of nature" (Prigogine and Stengers, 288). Whether he will move on to a physics that is more fully post-modern or post-classical, rooting time in something like Whitehead's "concrescence" or Bohm's "enfoldment," remains to be seen.

1.4. Physics and Pantemporalism

Thus far I have written as if physics could be perfectly compatible with a pantemporalist perspective. But is this so? What of the arguments of those who claim the opposite—that physics *entails* the rejection of the view of time outlined above, with its closed past, open future, and creative present? For example, Willard Quine has said that the principle of relativity "leaves no reasonable alternative to treating time as space-like" (Quine, 1960, 172). Costa de Beauregard has said:

There can no longer be any objective and essential (that is, not arbitrary) division of space-time between "events which have already occurred" and "events which have not yet occurred." . . . Relativity is a theory in which everything is "written" and where change is only relative to the perceptual mode of living beings.[39]

P. C. W. Davies states that:

relativity physics has shifted the moving present out from the superstructure of the universe, into the minds of human beings, where it belongs. . . . The four-dimensional space-time of physics makes no provision whatever for either a "present moment" or a "movement" of time. (Davies, 1976, 2f., 21)

Davies then quotes with approval Herman Weyl's statement that "the objective world simply *is*, it does not happen." Fritjof Capra says:

The relativistic theory of particle interactions shows thus a complete symmetry with regard to the direction of time. . . . This, then, is the full meaning of space-time in relativistic physics. Space and time are fully equivalent. . . . To get the right feeling for the relativistic world of particles, we must "forget the lapse of time." (Capra, 1975, 183–85)

A clearer example of the spatialization of time could not be wanted. First time and space are said to be equivalent; then this equivalence is taken to mean that time is eliminated. Capra then quotes with approval the famous statement of de Broglie (which is included among the epigraphs at the head of this essay).

However, it appears that the fallacy of misplaced concreteness has been committed again. The abstractness of the time in the abstract space-time of physical theory is forgotten, and it is treated as if it could be equated with the temporal structure of reality itself. As Whitehead says, "an abstraction is nothing else than the omission of part of the truth" (1966, 138). The same point was seen clearly by Mary F. Cleugh. In *Time and Its Importance in Modern Thought,* published in 1937, and still one of the best books on the topic, she asks the crucial question:

A fundamental feature of time as experienced is its irreversibility: is this really so, or is it merely an anthropomorphic prejudice, and is physics right in abstracting from this? (Cleugh, 1937, 49)

Her answer is that we need to distinguish between *legitimate* and *falsifying* abstraction and that the physicist's abstraction from time's irreversibility becomes a falsifying abstraction when it is taken to be a metaphysical truth:

It cannot be too often emphasized that physics is concerned with the measurement of time, rather than with the essentially metaphysical question as to its nature. . . . We must not believe that physical theories can ultimately solve the metaphysical problems that time raises, or that they have any *special* relevance to these problems. (51)

In support of the abstractness of the physicist's time, she quotes A. A. Merrill's article, "The *t* of Physics":

t, while created originally from our direct experience with real time, is subsequently handled in a way that has no relation to real time at all.[40]

She then concludes:

If it is claimed, whether openly or by implication, that the characteristics of *"t"* give a finally satisfactory account of time, that claim is unfounded. . . . The pliable, reversible *"t"* may be very useful and important in its own sphere, but its sphere is not that of metaphysics. (Cleugh, 1937, 50)

(Lest some be tempted to consider this type of point irrelevant, on the grounds that they are not interested in metaphysics, it should be emphasized that *any* attempt to state something about the *true nature* of time is "metaphysical," as the term is being used here.)

In an article entitled "Time Represented as Space," Nathaniel Lawrence argues similarly, addressing "the vanity which holds that the abstract considerations of material science provide an adequate framework for understanding our experience of temporality" (Lawrence, 1971, 123f.). Physics, he says, is a mode of practice focused upon measurement. For this purpose, it abstracts from time's passage, its additive (or cumulative) character, its absolute difference from spatiality, and its qualitative aspects. The result is time represented as space. There is justification for this abstract representation in the need for measurement. But there is also danger, especially because in certain respects physics has been so successful:

The great danger in these restricted enterprises is success. Success in one's own particular practice convinces him that he has got his hands on the primary reality. And therefore the more he will argue that other visions of reality are best tested by one's own particular discipline. (123)

But the truth, Lawrence says, is that

There is no mode of practice whose presuppositions are adequate for generating a total philosophy. . . . Measurement is almost as hopelessly partial as an approach to reality as is the marketing of peas. (123, 129)

Lawrence's position reflects that of Whitehead (1958, 27; 1958, 11), who stressed that we must distinguish between the authority of science to establish its own methodology and its competence (i.e., lack of it) to establish our ultimate categories of explanation.

In the same vein is the thought of Arthur Eddington, who points out that the abstractness of time in physics is not unique to it.

Physics has no concern with the feeling of "becoming" which we regard as inherently belonging to the nature of time, and it treats time merely as a symbol; but equally matter and all else in the physical world have been reduced to a shadowy symbolism. (Eddington, 1968, 22)

Like Whitehead, Eddington is trying to warn people not to equate these shadowy symbols, or abstractions, with the concrete realities of the world. With reference to relativity theory in particular, Eddington says:

Those who suspect that Einstein's theory is playing unjustifiable tricks with time should realize that it leaves entirely untouched that time succession of which we have intuitive knowledge, and confines itself to overhauling the artificial scheme of time which Römer first introduced into physics. (Eddington, 1968, 18)

Mary Cleugh comments on the widespread idea that relativity reduced time to a dimension of space.

In the new physics, when it is said that the dimensions of "space-time" are at right angles to each other, that does not mean . . . that time is somehow reduced to a dimension of space. It means precisely the contrary. (Cleugh, 1937, 69)

For support, she cites Einstein himself:

The non-divisibility of the four-dimensional continuum of events does not at all, however, involve the equivalence of the space co-ordinates with the time co-ordinate. On the contrary, we must remember that the time co-ordinate is defined physically wholly differently from the space co-ordinates. (Einstein, 1950, 31)

Likewise, in reply to Emile Meyerson's complaint against the tendency of many to reduce time to a fourth dimension of space (see introductory epigraph), Einstein wrote:

Meyerson rightly insists on the error of many expositions of Relativity which refer to the "spatialization of time." . . . The tendency he denounces, although often latent, is nonetheless real and profound in the mind of the physicist, as is unequivocally shown by the extravagances of the vulgarizers and even of many scientists in their expositions of Relativity.[41]

Phillip Frank also points out that the metaphysical "extravagances" regarding relativity theory are not only due to the "vulgarizers," but "have their origin in the insufficiently clear formulations which can be found in treatises of physics themselves" (Frank, 1976, 388). In the article entitled "Is the Future Already Here?" Frank's target is the attempt to justify from relativity theory the metaphysical view "that everything that happens is determined from all eternity, and that there is no development and nothing really new in the world" (387). As an example of an extravagant metaphysical claim based upon confusion, he cites (389) James Jeans's statement:

It is meaningless to speak of the facts which are apt to come . . . and it is futile to speak of trying to alter them, because, although they may be yet to come for us, they may already have come for others.[42]

Milič Čapek has devoted the most attention to this topic in recent times. He rejects as confused those inferences from relativity theory according to which the relativization of simultaneity destroys the objectivity of temporal order, so that events succeeding each other in one inertial system could appear in reverse order in another system (Čapek, 1976, 506, 507, 511). The truth is, he argues, that there is much absoluteness in relativity theory:

The irreversibility of the world lines has an *absolute* significance, independent of conventional choice of the system of reference. . . . Since the universe consists of the dynamical network of the irreversible causal lines, their irreversibility which remains absolute in the relativity theory is conferred to the universe as a whole. (514, 515)

Since (the causal) "before-after" relation is invariant in *all* systems, it follows that *in no frame of reference can my particular "here-now" appear simultaneous with any event of my causal future or with any event in my causal past.* (518f.)

The idea expressed in the above quotation from Jeans is categorically rejected by Čapek:

No event of my causal future can ever be contained in the causal past of any conceivably real observer. . . . *No event which has not yet happened in my present "here-now" system could possibly have happened in any other system.* (519)

Thus the virtualities of our future history which our earthly "now" separates from our causal past *remain potentialities for all contemporary observers in the universe.* Something which did not yet happen for us could not have happened "elsewhere" in the universe. (521)

Hence, whether one sees the "time" of relativity theory to be too abstract to be of direct relevance for the philosophical discussion of time, with Cleugh, Eddington, and Lawrence, or whether one sees relativity theory as having philosophical significance, with Čapek and Prigogine, there is good reason to question the counterintuitive ideas that some physicists, speaking as metaphysicians, have claimed to be warranted by it.

What about the claim that, according to elementary particle physics, particles can move backwards in time? This notion has come into circulation largely through the influence of Richard Feynman's diagrams and suggestions. Feynman suggested that, rather than interpreting certain interactions as the production and annihilation of particles, it is simpler and preferable to speak of some of the particles as going backwards in time. However, P. J. Zwart has argued convincingly that the original interpretation in terms of pair production and pair annihilation is much less problematic than the interpretation based on time reversal (Zwart, 1976, 155–59). Insofar as the idea of the production and annihilation of particles is thought to be problematic, since it suggests creation out of nothing, Milič Čapek's comment (1976, 517) is relevant:

It is not true that the process described above involves "creation from nothing" and "vanishing into nothing"; the pair of particles arises from electromagnetic radiation, into which it can be reconverted.

As pointed out earlier, Whitehead's suggestion that a particle is really a temporally-ordered society of momentary energy events makes the idea of the emergence and disappearance of "particles" from a background field of relatively chaotic energy events seem less counterintuitive than it would otherwise be.

Hence, there is no good reason to hold that any conventions in physics should lead us to think that time is unreal, or reversible, on the grounds that there are things that really "move backwards in time." Even if we had no other reason for rejecting this suggestion, the mere fact that no one can say what that conjunction of words *means* should give us pause.

I close this section by pointing out that Prigogine emphatically rejects the view that quantum and relativity theory should give any aid and comfort to eternalists, even if they could have once been so interpreted. As he sees it, quantum theory now describes the transformation of unstable particles, and general relativity is now seen to describe the thermal history of the universe (Prigogine and Stengers, 1984, 9). Although Einstein's own 1917 picture of general relativity presented a static, timeless, Spinozistic vision of the universe, it was soon shown that there are time-dependent solutions to his cosmological equations (215). Finally, instead of regarding the joining of space and time in relativity as implying the spatialization of time, Prigogine agrees with those, such as Bergson and Čapek (1983), who say that it is more accurate to speak of the temporalization of space (17).

1.5. Importance of the Topic

Why is the topic of this volume important? That is, why is it important whether time, with its asymmetry, becoming, and irreversibility, is ultimately real or illusory? And why is it important to get clear about the relevance of physics to time and of time to physics? I shall conclude this introductory essay with some reflections on these questions—still from a perspective that is deeply shaped by process philosophy. I shall organize this discussion in terms of six reasons for the relevance of the topic.

(1) The importance of a temporalized physics for overcoming the dichotomy that has existed in the modern world between the natural sciences and the humanities has already been discussed in relation to Prigogine's thought; it need not be elaborated upon further.

(2) However, there is also the problem of the dichotomy *within* the natural sciences, between fundamental physics and everything else. For, as Stephen Toulmin and June Goodfield (1965, 247) point out in *The Discovery of Time*,

The physical sciences had stood aside from the historical revolution which transformed the rest of natural science, taking it as axiomatic that certain aspects of the world remained fixed and permanent throughout all other natural changes; and though . . . the list of these timeless entities . . . is much shorter than it was in 1700, the existence of unchanging physical laws, at least, is still regarded as one enduring aspect of the natural world.

Like Prigogine, Toulmin and Goodfield see that this idea of physics was conditioned by theological ideas: "since God Himself was regarded as changeless and eternal, it was presumed that the 'laws of nature' which were an expression of His will were correspondingly fixed in their form" (264). They see it to be "still an open question whether physics will ever become a completely historical science" (247).

The outstanding question now, is whether the laws of Nature themselves—the last *a-historical* feature of the physicists' world-picture—will in their turn prove to be subject to the flux of time. (250)

This idea had been developed by Whitehead, who suggested that we look upon the so-called laws of nature as simply the most long-lasting and widespread "habits of nature," i.e., the habits taken by the most elementary processes constituting our particular universe (1938, 154). This means that "natural laws" and "sociological laws" would be different only in degree, not in kind. They would be more or less universal patterns of behavior which tend to be transmitted from generation to generation, but which can also undergo more or less gradual changes. In Whitehead's view the so-called elementary particles, such as electrons, protons, and neutrons, which emerged at some point in the past (1958, 24f.), are really temporally-ordered *societies* of momentary energetic events that have been stable enough to endure for billions of years. They did not emerge from absolute nothingness, but from a realm of chaotic events. Surely it is hard to think of the laws that describe the behavior of all such "particles" as preexisting them. Accordingly, if the idea that the "elementary particles" of nature have evolved is accepted, it is most natural to think of the laws of physics as themselves having developed at some period in the past. And, once this thought is accepted, it is natural to assume that their validity will be for a limited duration (1926, 65), i.e., that they will evolve some time in the future, when the entities whose behavior they describe evolve.

The idea that electrons, protons, and neutrons have evolved historically from a relatively chaotic field of events has been given additional support by more recent developments in elementary-particle physics, as now there are considered to be over a hundred forms of such "particles." But most of these forms exist only very briefly, many for less than a billionth of a second. This constant appearance of other particles from the background field and disappearance back into it supports the notion that electrons, protons, and neutrons also emerged from such a background and are simply much more stable forms of temporally-ordered societies of momentary energetic events—but not so stable that they will not eventually disappear.

Also, we have learned that the various forms of elementary particles, including photons, can be transformed into other types. This further supports the notion that all the basic forms of matter are only partially stable "societies" of momentary events, which have evolved into their present forms and which may presently be undergoing some slow and subtle evolution. Again, if what we call "matter" has itself evolved, is still evolving, and will some day no longer exist, how can it make sense to speak of the "laws of matter" as themselves timeless? Must we not assume that these laws evolved correlatively with the entities whose behavior they describe, just as biological laws evolved along with the living things whose behavior they describe and as sociological laws evolved with the types of human beings whose behavior they describe? If this idea is accepted, then time, with its difference between past, present, and future, its "moving now," and its irreversibility in principle, will have to be recognized as applying to the fundamental laws of physics, for it will thereby be recognized that the so-called fundamental laws themselves have a history, that their importance in the actual world had a beginning and will have an end.[43]

When this occurs, then physics will finally, with the rest of the physical sciences, have discovered time, and one more absolute dichotomy which prevents our developing a unified world view on the basis of the evolutionary paradigm will have been overcome.[44]

(3) The dichotomy between temporal and nontemporal is correlative with a dichotomy between freedom as real and as illusory, for without becoming, in which the present involves turning potentiality into actuality, there can be no freedom (as stressed in Frederick Ferré's concluding essay). This dichotomy has led to tragic divisions within human experience. On the one hand, the natural sciences in general, and physics in particular, have gained such authority in our culture that it is very difficult to believe something wholeheartedly if physical theory seems to contradict it. On the other hand, it is also impossible, as I have suggested, consistently to deny those "hard-core common-sense" notions which all people presuppose in practice, as evidenced by their behavior. "Freedom," the idea that we are to some degree free to determine our own responses to events, is one of those notions.

It implies the partial openness of the future, and yet many have taken physics to imply that the "future" is already as fully determinate and actual as the past; see, for example, the epigraphs at the head of this essay by de Broglie, Weyl, Grünbaum, and Einstein. Hence, people have been torn between contradictory ideas. Einstein himself proves an example. In one of the epigraphs, he says that freedom is an illusion, but then adds tellingly: "even if a stubborn one." His own life displays a marked contrast with his writings. In his writings he portrays a Spinozistic universe, in which all things are eternally determined (Einstein, 1954, 48–50, 55–56); but he devoted much of his energy to the passionate quest to avert atomic war—showing by his practice that he knew that our fate is not already in the stars (121–67). Finally, Rudolph Carnap reports that, near the end of his life, Einstein was deeply worried by the awareness that there is something about the present Now that makes it essentially different from the past and the future, but that this is a difference which "does not and cannot occur within physics" (Carnap, 1963, 37f). (For more on Einstein's views, see Popper, 1982, II, 2–3n, 89–92.)

(4) The tension between fate and freedom is only one aspect of the dichotomy between the sciences and the humanities. As Prigogine stresses, the dichotomy of the temporal and nontemporal is at the root of a more general opposition between humanity and nature, which involves the clash between the qualitative and the purely quantitative, between the active and the passive. The idea that the world of nature is a passive realm, devoid of aesthetic qualities and intrinsic value, has led to a devaluation of these values in modern life. Hence, the issue of the ultimate status of time is part of the overall problem of modern thought, which has, among other things, contributed significantly to the ecological crisis of our time.

(5) Another issue is the tension between modern physics and traditional religions. In much semipopular writing, the impression is given that traditional Christian thought has affirmed the asymmetry, and hence ultimate reality, of time, whereas modern physics gives us this great, new, liberating idea that the relations of the present to the past and to the future are symmetrical, so that time is unreal; and that, on this point, modern physics agrees with Buddhism and other forms of mystical and perennial philosophy.

However, as the song puts it: "It ain't necessarily so." On the one hand, traditional Christian theology has portrayed God as eternally and immutably knowing the world. This means that what is future to us has to be as present to God as what is past and present to us. Indeed, God was said to know the entire history of the world in one simultaneous now (*simul nunc*). It was only with recent process philosophies and theologies that we have Christian theological systems that genuinely allow for the reality of time. It is the reality of time, *not* its denial, that is the new idea in Western philosophical and theological thought.[44a]

On the other hand, Buddhism by no means unambiguously affirms the symmetry of past and future. The *mutual* interfusion of past, present, and future, which Fritjof Capra stresses in *The Tao of Physics,* was mainly advocated by one school of Chinese Buddhism, Hua-Yen, [45] which then influenced some branches of Japanese Buddhism, including Zen.[46] There are Buddhists who strongly disagree with the idea of temporal symmetry and who see it as antithetical to Buddhist religion.[47] For one thing, this idea that the future is as influential upon the present as is the past stands in strong tension with the idea of karma, which pervades all Buddhist thought, according to which the causal influences upon one arise from the past, but that one can so act in the present as to become liberated from bad karmic influences.

Finally, Ken Wilber, one of the leading theoreticians of transpersonal psychology, which is based on mysticism and the so-called perennial philosophy, emphatically rejects the idea that the relation of the present to the future is no different from that to the past. In fact, Wilber endorses Whitehead's views on this subject: an event prehends only its predecessors, not its descendants; Christopher Columbus affects us, but we do not affect him (Wilber, 1982, 286–88). Evidently neither Buddhism, nor mystical experience, nor the "perennial philosophy" can be appealed to as an unambiguous witness to the mutual interpenetration of past, present, and future and hence to the ultimate unreality of temporal distinctions. Just as it is not a question of what "physics" says, but of what individual physicists or philosophers of physics say (as pointed out in the introductory epigraph by Gale), so it is not a simple question of what "mystical experience" or "enlightenment" or "Buddhism" says, but of what is said by individual *interpreters,* with their particular histories, biases, preconceptions, and selective attentions. As Whitehead says, if you want uninterpreted experience, ask a stone for its autobiography (1978, 15).

Wilber also makes reference to so-called precognitive experiences, indicating that, however they are to be interpreted, they must not be interpreted so as to deny the reality of free will in the present (Wilber, 1982, 287). This is a point upon which I want to enlarge. We could probably not exaggerate the extent to which ideas of the ultimate unreality of time in religious thought have been based on such experiences. Such experiences—in which a vision turns out to correspond with a later event—have been widespread, especially in contexts out of which religious writings arose. The most common interpretations of these experiences have been these: (a) God reveals to a prophet what God "already" knows is going to happen; (b) in meditation, one comes into contact with a level of existence in which all things, which in ordinary consciousness seem to be distinguishable as past, present, and future, exist simultaneously in an "eternal now"; or (c) a future event causes the present vision. All three of these interpretations imply that that realm which appears to be future from the standpoint of any present "now" is in reality as filled with fully determinate events as is the past. Hence, there is no categorical difference between past and future, so it is purely arbitrary, as Bertrand Russell says, that we remember only the past and not the future. So-called precognitive experiences are widely taken as evidence that this arbitrariness is sometimes overcome, as some people "remember" events in the future.

However, these are all tendentious interpretations. Purely phenomenologically, what is experienced is (i) a vision followed by (ii) an event that corresponds closely to the vision (too closely to be dismissed as mere coincidence). That is, there is nothing in the experience as such that dictates that when the vision occurred the event already existed or that the (later) event caused the (prior) vision. It would be equally compatible with the evidence to say that *the vision caused the subsequent event to occur.* In fact, there are thoughtful interpretations of so-called precognitive experiences that employ this notion (Eisenbud, 1956, 23–25; Ebon, 1968, 224; Tanagras, 1967). However, most so-called precognitive experiences can be explained within a pantemporalist framework without resort to such a drastic idea. For example, most so-called precognitions of disasters can be accounted for in terms of unconscious clairvoyant knowledge of present structural defects (e.g., a leaky ship, a cracked wing, a short circuit in electric wiring, a weak heart, a cancerous growth), plus unconscious inference of the probable outcome, resulting in a dreamlike vision rising to conscious awareness. Most other so-called[48] precognitions can be accounted for by unconscious telepathic knowledge (of other people's knowledge, feelings, or intentions—perhaps unconscious on their part), plus unconscious inference. For other cases not easily explainable in one of these three ways, there are several other ways that do not require reverse causation, or an already actual future, or anything else implying the unreality of time.[49] So, again, there is nothing about these special experiences that should encourage us to deny the temporality of reality, which is suggested by the overwhelming majority of our experience, secular and religious alike.

To sum up the main point of this discussion: in response to the claim of Prigogine and process philosophers that a nontemporal interpretation of physics isolates it from the rest of human culture, some might reply, "Not entirely. Certainly, it isolates physics from conventional interpretations of experience and reality, including conventional Western religious thought, but it puts physics in harmony with those experiences in which human beings have transcended conventional consciousness and plunged more deeply into the nature of reality. On the basis of those experiences have arisen interpretations of reality that are remarkably similar to those arising from modern quantum and relativity physics." Over against this view, my argument has been that a nontemporal view of reality is not unambiguously supported by any kind of nonscientific experience—ordinary, religious, or parapsychological. Accordingly, there is good reason to hold to the original point: nontemporal interpretations of physics will continue to stand in contradiction with the rest of human experience and the best interpretations of reality based on it.

(6) There is special reason for focusing upon the question of physics and the ultimate significance of time at the present moment. There has recently been a spate of popular interpretations of physics that have reached wide audiences. These books have spread quite widely the idea that the authority of physics supports the notion that time is unreal. Reference has already been made to Fritjof Capra's immensely popular *Tao of Physics*. Gary Zukav's *Dancing Wu Li Masters* has also been a best-seller. In it, the reader learns that the preferred interpretation of quantum field theory is to speak of anti-particles as particles traveling backwards in time (Zukav, 1979, 236, 237). The reader encounters a quotation from de Broglie and learns that it is an illusion that events "develop" in time (238). Zukav's moral is similar to Capra's: since the flow of time has no meaning at the quantum level, and consciousness itself may be a quantum process, we can perhaps experience timelessness (240). Fred Alan Wolf, whose *Taking the Quantum Leap* was quite popular, now tells us in *Star Wave* not only that time is unreal (1984, vii, 230), but that it is in our power to create the past (vii, 109, 110, 187), so that what "really happened" is determined by a majority vote (230, 326). Accordingly, it is within our power to determine whether the Nazi Holocaust occurred—not just epistemically, but ontically: we could make it so that it really had not happened.[50]

These metaphysical speculations, which claim to be warranted by the "new physics," are particularly egregious examples of the havoc that can be created in people's thinking by the fallacy of misplaced concreteness. The interpretation of quantum physics from which this conclusion of the unreality of time is drawn is one which *methodologically abstracts* from the question of what is really going on in the microscopic realm. It rejects the attempt at a realistic interpretation in favor of a phenomenalistic one. That is, it restricts itself to formulae that successfully correlate the macroscopic observations, i.e., the instrument readings. (Bohm's long interest in "hidden variables" was less motivated by the desire to confirm Einstein's and de Broglie's deterministic vision than to find a theory that would describe the real processes going on.) But then interpreters forget the abstraction from reality and start applying details of the intentionally phenomenalistic account to the operations of the real world. In Wolf's case, the crucial idea is that no quantum reality has determinate status until it is measured, it being the measurement that is said to "collapse the wave function." The paradoxical implication, drawn unflinchingly by some interpreters, is that there can be no determinate reality apart from human observers, so that even what we normally think of as the distant past is dependent upon present observation.[51] This would mean that the entire theory of evolution would have to be abandoned, for that theory (which virtually all of contemporary science presupposes) assumes that quite determinate processes did occur for billions of years prior to the rise of human (or humanlike) observers. Such a conclusion should alert us that something is amiss somewhere.

However, there is little danger that masses of laypeople are going to be led to believe that the past is under their control and hence to begin taking their decisions less seriously on the grounds that, if they do not like the results (e.g., a nuclear war) they can simply revise the past so that those events did not occur (assuming that there are any of us left to carry out the revision). Our sense of the immutability of the past is too strong to be shaken, even if we be told that we are rejecting the authority of physics. After all, classical theists, never shy about affirming divine omnipotence, would not even ascribe to God the power to undo the past. Accordingly, affirming the symmetry of past and future[52] usually means regarding not the past as open but the future as closed. The introductory epigraphs by de Broglie, Weyl, Gold, Grünbaum, Russell, and Einstein all reflect this view. The effect these ideas can have upon the intelligent layman, even with some scientific training, is illustrated by a book by Larry Dossey, M.D., entitled *Space, Time and Medicine.*

Dossey begins his chapter on "Modern Time" with the introductory epigraphs by Gold and de Broglie, after another by Bertrand Russell, which reads:

A truer image of the world . . . is obtained by picturing things as entering into the stream of time from an eternal world outside, rather than from a view which regards time as the devouring tyrant of all that is. (Russell, 1917, 21)

After introducing the idea of creativity, Dossey draws these conclusions:

This concept of creativity as a feature of a timeless, eternal world is hard to swallow, especially if one wishes to see his creative act as a literal making of some new thing. But in the modern context of a nonlinear time, as de Broglie states, events exist prior to us in time. We create nothing, since all things exist already. (Dossey, 1982, 34)

As the beginning of that statement shows, Dossey is keenly aware that this view of time goes directly counter to our ordinary notion of time—what I have called our "hard-core commonsense" view. But, illustrating my earlier point about the authority generally granted to physicists in our culture, he shows no hesitation as to which view to adopt: "The view of time from modern physics tells us our ordinary notions of time are wrong" (41). After quoting the famous statement from Weyl, he says: "This view is an affront to common sense. . . . [But we] must assimilate it," for "we cannot ignore what modern physical science has revealed to us about the nature of time" (152, 153). He justifies this rejection of "common sense" by quoting Einstein's statement that "common sense is merely the deposit of prejudices laid down in the human mind before the age of eighteen." This is probably true of most of those notions that are usually classed as "common sense" and that I have called "soft-core common sense." But is there not an important difference between those culturally conditioned ideas, out of which we *can* be educated, and the true (i.e., hard-core) commonsense ideas which belong to the *sensus communis* of humanity and which we all inevitably presuppose in practice? Do we not have good reason to have more confidence in them than in any ideas from which their contraries can be deduced? And does it not belong to our hard-core common sense that the future is not totally determined? If so, then we should take those ideas as the criteria for evaluating the alleged "revelations" from modern physics, instead of the other way around.

The power of the physicalist conception, that if something cannot be detected by the methods of science then it does not exist, is shown by the fact that Dossey repeatedly cites with approval the statement of P. C. W. Davies that no physical experiment has ever been performed that detects the passage of time (Dossey, 151). Dossey then provides a summary of an "incontrovertible lesson" from modern physics, which amounts to a complete endorsement of Davies's views:

The notion that time flows in a one-way fashion is a property of our consciousness. It is a subjective phenomenon and is a property that simply cannot be demonstrated in the natural world. This is an incontrovertible lesson from modern science. . . . A flowing time belongs to our mind, not to nature. We serially perceive events that simply "are". . . . (151)

Dossey's concerns are practical. He believes that this view can help people overcome the anxiety, and related health problems, caused by looking upon time as "the devouring tyrant" (146). He suggests that we learn from modern physics that time is a chimera and that "length of life is meaningless for the reason that passage of linear time does not occur" (145). His conclusions are well-intentioned, but whereas it may be healthy, in some sense, to apply to oneself the idea that "length of life is meaningless," to take this as a generally true statement, and to apply it to one's enemies, or to humanity as a whole, could be extremely dangerous—especially in this nuclear age.

So, if time is real, and if each moment is an opportunity to decide which among a variety of potential values will become actual, thereby conditioning the entire future of the universe for good or for ill, then the idea that time is not real in this sense is a false and dangerous one. It will lead us to take our enormous power for decisionmaking, and hence value-realization, less seriously than we should. The same is true for the idea that the past can be undone.

Given the tremendous authority granted to physics and hence physicists in our culture, and the importance that our ideas about time have in our attitudes toward life and the way we live it, it is crucial that we become clearer about what contemporary physics does and does not entail about the nature and ultimate significance of time. It is my hope that this volume will make a helpful contribution toward this end. In this introductory essay, I have pointed out various paradoxes about time that have plagued human thought, especially in the modern period. I have sought to give reasons for taking seriously in each case the suggestion of Whitehead that "the paradox only arises because we have mistaken our abstraction[s] for concrete realities" (1926, 80f.)

Notes

1. This is the behavior of the K meson; see Whitrow (1980, 355f.); Davies (1976, 175f.).

2. Some writers unfortunately use the term *asymmetry* for anisotropy. I use *asymmetry* to mean that the present's relation to the past is categorically different from its relation to the future. When P. C. W. Davies speaks of "the

physics of time asymmetry" he does not have asymmetry in this sense in mind, but only what others (e.g., Grünbaum, 1967) less ambiguously call "anisotropy." For example, Davies is at pains to distinguish "the real, physical asymmetry" from "the controversial phenomenon of psychological time," which divides time into the past and the future separated by a now (Davies, 1976, 3, 20–22).

3. "Nothing yet discovered in nature requires individual atoms to experience time asymmetry, the very essence of which is the collective quality of complex systems, like life itself" (Davies, 1976, 4).

4. Richard Feynman says that "irreversibility is caused by the general accidents of life. . . . It is not against the laws of physics that the molecules bounce around so that they separate. It is just unlikely. It would never happen in a million years. . . . Things are irreversible only in a sense that going one way is likely, but going the other way, although it is possible and is according to the laws of physics, would not happen in a million years" (Feynman, 1965, 112). Writers who so derive the irreversibility of time from entropy hold that our memory is irreversible only because we are ourselves complex systems exemplifying entropic principles (Feynman, 1965, 121; Davies, 1976, 19–22). Accordingly, if thermodynamic systems were to reverse directions, our memory would presumably reverse too, with the result that we would remember the future but only predict the past. One should consider this conclusion a *reductio ad absurdum* of the postition. Percy Bridgman agrees (see epigraph at head of this introduction; Bridgman, 1955, 251); so does P. J. Zwart, who considers it absurd to think the temporal order is based on the entropic order (Zwart, 1973, 144). He says: "One thing is quite inconceivable: that we could perceive a later event before an earlier one. . . . This kind of proposition is self-contradictory, that is to say, time reversal in this sense is *logically* impossible. Reversal of the entropic ordering is not logically impossible, however, though highly unlikely. But even if it occurred, we should only have to adjust our *laws,* not our *concepts*" (Zwart, 1973, 145). Likewise K. G. Denbigh: "Mental processes display irreversibility of a kind not shown by physical processes— that is, in the sense that it is *not conceivable* that they could ever occur in the reverse temporal sequence" (Denbigh, 1975, 39). He points out that, if we had settled upon some physical process as a standard for deciding the before-after relationship—e.g., an apple falling from a tree to the ground—and then found that the event sequence went in reverse order, we would simply decide that the "standard" was falacious. "In other words, is it not the case that *what we really mean* by the *before-after* relation is the relation as it is offered by consciousness?" (Denbigh, 1975, 41). My only difference with Denbigh here is that, from the viewpoint of process philosophy, the notion that "physical processes" are reversible results from conceiving of them at an abstract level and hence in effect conceiving of them as *types* of processes. If we took experiences, or what he calls "mental processes," at the same level of abstractness, they also would be considered reversible.

5. Richard M. Gale (1968, 107) says that it is an "analytic truth that a present cause cannot have a past effect." Process philosophers agree, while rooting this truth of language in the structure of experience and finally of actuality in general.

6. Even Einstein evidently came to believe this; see the discussion in section 1.4 in the text below.

7. See, for example, H. Bondi (1952) and H. Reichenbach (1956, 269).

8. Henry Mehlberg (1980, 202) asks how we can reconcile the fact that our knowledge of the past differs so greatly from our anticipation of the future with the fact, vouchsafed by science, that time is isotropic, having no intrinsic difference between past and future. He suggests that "the only possible answer is the realization that temporal words like 'past,' 'future,' 'planning,' etc. have no independent meaning," but depend upon the usage of one's linguistic community. "What I remember belongs to my past, by definition, and what I desire or am planning for, belongs to my future, by definition equally. But this need not prevent somebody else from desiring what I remember and, thus, from having his future overlapping with my past." Should such a conclusion not lead one to suspect that something is amiss?

9. Adolf Grünbaum (1967, 152f.) says that "the transient now with respect to which the distinction between the past and the future of common sense and psychological time acquires meaning has no relevance at all to the time of physical events." Accordingly, "coming *into* being (or 'becoming') is *not* a property of physical events themselves but only of human or other conscious awareness of these events" (154).

10. The foregoing account was meant as an explication of those features that are largely agreed upon by authors from diverse perspectives, not as an adequate account of time as we experience it. For it excludes the crucial feature, discussed below in the explication of Whitehead's view, that time as experienced always involves conformity to the past (and anticipation of the future's conformity to the present). It is this conformity to (and in fact inclusion of) the past that is the strongest reason for the irreversibility of time. The theory of time including this feature is a causal as well as relational theory. My one major difference from P. J. Zwart arises here, in that Zwart (1973, 131–33, 144) holds that time is to be understood relationally but not causally, as he regards the temporal order as more fundamental than the causal. However, the difference seems to arise because he thinks of events, causes, and effects as abstractions, i.e., *types* of events. This is clear in his argument that if "we should suddenly perceive that all the events we used to call the effects of other events were preceding these latter events, it is inconceivable that we should adjust the temporal order to the causal order, instead of the reverse" (144). Apparently for this reason he can say: "If temporal order could be reduced to causal order 'A precedes B' would have to imply 'A is the cause of B' which, of course, is not at all true" (144). But if we understand A and B not as types of events but as concrete, singular events, and also that by calling A a "cause" of B we mean that it was an event without which B could not have been precisely what it

was, then Zwart could perhaps accept the equation of causal and temporal order.

11. As K. G. Denbigh says, the view that the universe is symmetric temporally would lead to the idea that the "happening" or "taking place" of events must be thought of as a sort of illusion of consciousness that has no obective counterpart (1975, 7). In the introduction to a symposium on time, Thomas Gold (1967, 2) asks whether the notion of a flow of time, which we get from introspection, is "only a deception of a biological sort which ought to have no place in physics." Later he reveals that this is indeed what he holds: "We ought to eliminate this flow idea from the real picture [of the world], but before we can eliminate it we ought to understand how it arises. We should understand that there can be a self-consistent set of rules that would give a beast this kind of phoney picture of time" (182). He closes the volume with a hope for "some new physical theory" that will give a "better description of nature, one that is less dependent on our subjective notions" (243). P. C. Davies (1976, 22) speaks of "the apparently illusory forward flow of psychological time." His reductionistic position is especially revealed in a statement following his observation that, although the laws of physics do not provide a time asymmetry, the world *as a matter of fact* is asymmetric (i.e., anisotropic; see note 2) in time. He concludes from this that the world's anisotropy is hence "extrinsic" or "factlike" rather than "intrinsic" or "lawlike." These equations show that, for him, nothing is *intrinsic* to the world unless it is contained in the fundamental *laws* of physics. My object is not to criticize those who hold these positions but to show, first, that many people *have* adopted such positions and, second, to suggest a framework in which both their concerns and a more adequate view of time can be combined.

12. See note 4.

13. For example, Mary F. Cleugh (see the first epigraph by her) and Robert S. Brumbaugh (1984, 6,10): "The study of time is a metaphysical enterprise, not a physical one. . . . The nature of time is a *philosophic* problem. In physics, time is an undefined basic notion."

14. Whitehead describes the fallacy as "mistaking the abstract for the concrete" (1926, 74f.) One major form of it is the description of actual things as if they exemplified "simple location," i.e., as if they could adequately be described as simply being at one place in space-time, apart from any essential reference to other regions of space and other durations of time (1926, 74f., 84).

15. "The reality is the process" (1926, 106); "the very essence of real actuality—that is, of the completely real—is *process* (1933, 354).

16. In this volume; and Čapek (1976, xxxiv–xxxv).

17. I have focused on this aspect of process thought in relation to the problem of evil in Griffin (1981).

18. See notes 8 and 11, and the first epigraph by Einstein, above.

19. I have put "elementary" in scare quotes to indicate distance from the notion of "elementarism," which is discussed in Bjelland's essay in this volume.

Elementarism implies that there are fundamental particles out of which more complex things are created but which themselves undergo no essential changes by becoming part of this new environment. This is exactly the notion of autonomous bits of matter with "simple location," which process philosophy rejects as a fallacy based on misplaced concreteness (see note 14). The term "elementary particles" also suggests that these entities are ultimate, i.e., that they do not require explanation, and are not thought to have evolved out of something more ultimate. Process philosophers share with Karl Popper (1982, III, 139, 158) the rejection of this notion of elementary particles.

20. For Whitehead, the "metaphysical rule of evidence" should be "that we must bow to those presumptions which, in despite of criticism, we still employ for the regulation of our lives" (1978, 151). It is these presumptions, which are inevitably presupposed in practice (Whitehead, 1978, 13), that I call our "hard-core commonsense notions." The chief problem with the so-called empiricist philosophies of Locke and Hume and their followers was that they did not take *these* empirical facts as basic but instead began with certain *a priori* dogmas about experience from which they either deduced the falsity of these hard-core commonsense notions (13, 146), or at best added them as presuppositions of *practice* as supplements to their metaphysical or epistemological *theory* (133f., 153, 156).

21. Although Richard Gale (1968, 103f.) does not explicitly distinguish between hard-core and soft-core commonsense, his position is very similar to mine. In his words, "many of our fundamental concepts, such as causality, action, deliberation, choice, intention, memory, knowledge, truth, possibility and identification, logically presuppose an asymmetry between the past and future. . . . These logical asymmetries interlock . . . such that one of them cannot be jettisoned without giving up the others as well. Together these asymmetries form the basis of our common-sense conceptual system." Karl Popper also points out that a major reason for rejecting a symmetrical view of past and future is that it conflicts with our common sense, which regards the past as closed but the future as not completely fixed. He says: "We should not be swayed by our theories to give up common sense to easily." But, failing to distinguish between soft-core and hard-core common sense, he does not consider common sense the ultimate arbiter, and hence *would* be willing to give it up in the face of a good scientific theory supporting the symmetrical view (1982, II, 3n., 55–56).

22. This is Roderick Chisholm's way of stating the "commonsense" approach: "I assume that we should be guided in philosophy by those propositions we all do presuppose in our ordinary activity. . . . Any philosophical theory which is inconsistent with any of these data is *prima facie* suspect" (Chisholm, 1976, 15, 18).

23. See Thomas Gold's statement cited in note 11.

24. Čapek (1976, xlvii); see also G. J. Whitrow (1980, 350).

25. This would be one of my main reasons for rejecting the view of Robert Brumbaugh (1984, 9, 11, 125, 131, 138, 139), according to which there are

four kinds of time in the world. I agree with the statement of G. J. Whitrow with which he distinguished his own position from that of J. T. Fraser (1975), which is similar to Brumbaugh's: "at all levels time is essentially the same, although certain aspects of it become increasingly significant the more complex the nature of the particular object or system studied" (Whitrow, 1980, 375). See also the discussion of Karl Popper's position in note 30a.

26. See note 14.

27. "[The] Kantian doctrine accepts Hume's naïve presupposition of 'simple occurrence'" . . . for the mere data. I have elsewhere called it the assumption of 'simple location,' by way of applying it to space as well as to time. I directly deny this doctrine of 'simple occurrence.' There is nothing which 'simply happens.' Such a belief is the baseless doctrine of time as 'pure succession.' . . . Pure succession of time is merely an abstract from the fundamental relationship of conformation" (Whitehead, 1958, 38).

28. Whitehead (1966, 138). Whitehead uses the term *matter* to refer to anything with the property of simple location (1926, 72). He does not think that matter in this sense is a total fiction; rather, he holds that "by a process of constructive abstraction we can arrive at abstractions which are the simply-located bits of material" (1926, 85). Again, the error is taking the intellect's abstraction from actuality to correspond, without essential loss, to the concrete actuality itself.

29. To my knowledge, Whitehead never used the term *temporally-ordered society*. Instead, he spoke of enduring objects with purely temporal or serial order as having "personal" order (1933, 259, 263; 1978, 35). However, he himself pointed out that the term *person* has connotations of consciousness (1978, 35), whereas his societies with personal order need not even be living (1933, 264), and indeed most of them—e.g., electrons, protons, atoms, molecules—are not. Hence I find the term *temporally-ordered society* preferable.

30. For criticisms of Grünbaum's thought on time, see Milič Čapek (1961, 377); and Frederick Ferré (1972).

30a. Karl Popper's thought provides an interesting counter example to the general correlation, since he is a dualist in regard to the mind-matter issue but not in regard to time. He agrees with process philosophers that the defense of realism requires the affirmation of the reality of time (1982, II, 3n.), that time's arrow is not derivative from entropy, that in any case there is more to time's asymmetry than its arrow, and that this asymmetry is incompatible with determinism (1982, II, 55–56). He grounds indeterminism and hence the non-derivative reality of time in the notion that all material particles realize and are "propensities": each thing is a realization of a prior potentiality and in turn a potentiality for another becoming (1982, III, 205, 207). This way of affirming the ultimate reality of time, and hence the temporality of physics, is very similar to the Whiteheadian notion that each actual entity is first a partially self-determining actualization of potentialities proffered it by past actual occasions and then in turn provides potentialities for future actualizations. However, although Popper attributes consciousness to amoebae, and an elementary memory to crystals and even to DNA molecules, he insists that atoms and subatomic particles are totally devoid of experience (1977, 29, 71). In his terminology, World I, which consists of purely material entities, existed prior to World II, which consists of psychological qualities (1982, 114–17). World I objects seem to have ontological as well as temporal priority: Popper shares "with old-fashioned materialists the view that . . . solid material bodies are the paradigms of reality" (1977, 10; cf. 1982, II, 116, 117). But this raises two questions: The first question is the traditional one: how can non-experiencing matter interact with experiencing mind? Popper simply ignores this problem, saying that "complete understanding, like complete knowledge, is unlikely to be achieved" (1977, 37). That is true, but there is an enormous difference between lacking complete understanding of how something occurs and affirming a position that makes the occurrence seem impossible in principle. The second question is based on the fact that "propensity," as Popper recognizes (1982, III, 209), is a psychological term. How are we to think of "solid material bodies" as having "propensities" even though they have nothing even remotely analogous to experience? The only statement I have found relating to this question is that the propensities are, like forces, "properties not so much of the particles as of the total physical situation; they are, like forces, relational properties" (1982, III, 127). But I do not see how this helps. Popper himself would evidently agree, since he says: "We take indeterminism as a cosmological fact which we do not attempt to explain" (1982, III, 181). Both of these problems are avoided if we assume that, below cells and DNA molecules, experience simply becomes less and less complex rather than disappearing altogether. Hence, I maintain my thesis that pantemporalism can be consistently defended only on the basis of panexperientialism.

31. Whitehead's is an "epochal" theory of time, meaning that time is derivative from the succession of actual events, each of which has a finite (i.e., noninstantaneous) duration. Although there *is* a space-time *continuum,* this is a structure of *potentiality.* The temporal process is not *in fact* continuous; it is not constituted by a series of instants. For a defense of Whitehead's position against criticisms, see David Sipfle (1971).

32. G. N. Lewis, who was perhaps the strongest advocate of time symmetry of all time, saw this point clearly. He saw the importance of conceiving of matter as strictly atomic, in the sense of being made of discrete particles that in no sense entered into each other, for if two things were allowed to diffuse into each other, "such a diffusion would in principle be a phenomenon which by no physical means could be reversed." But with discrete particles, the recurrence of every particular state will eventually come about. See Lewis (1930, 571) and the discussion in Roger Steuwer (1975).

33. See the discussion and quotations in Floyd Matson (1966, 38–56).

34. That H. Feigl's identism was really a dualism is evident. After saying that, instead of two types of events, "we have only one reality which is represented in two different conceptual systems," he makes this qualifying explanation: "on the one hand, that of physics and, on the other hand, where

applicable (in my opinion *only to an extremely small part of the world*) that of phenomenological psychology" (1960, 33; emphasis added). In other words, all events have physical qualities, but only a few have psychical qualities. Feigl says this explicitly: whereas "at most something very much less than a psyche is ascribed to plants or lower animals" (which implies no dualism), "nothing in the least like a psyche is ascribed to lifeless matter" (which does imply dualism).

35. Jacques Monod (1972, 21) has called this the "postulate of objectivity" and proclaimed it to be the hallmark of science itself. However, this is to dictate either that psychology must be behavioristic, eschewing all references to purpose, feeling, and the like, or that it cannot be a science. The same would be true of ethology. Do we really want to *define* science in such a way that it is forever excluded from speaking about subjectivity? At first glance, to say that science presupposes the "postulate of objectivity" seems unexceptional: *of course* science is to be objective! But this is to assume only one of the two meanings of this ambiguous term. That is, as long as the term is taken to refer to a *method of approach,* the postulate is a commonplace. But the term can also refer to the *categories used to describe the objects of study.* Monod takes the postulate of objectivity to include both meanings, with the result that science can use only categories of objectivity (e.g., mass, energy) as opposed to categories applicable to subjectivity (e.g., feeling, purpose). Science is thereby limited to treating all things as if they were objects in the metaphysical sense, i.e., objects as opposed to subjects. This limitation is arbitrary and is really nothing but an attempt to enshrine a particular metaphysics (materialistic or dualistic) into the permanent methodology of science. But, presuming that it is possible to be objective (methodologically) about subjectivity (as phenomenology and non-behavioristic psychology maintain), should we not define science in such a way that it is able to deal with whatever kinds of things the world contains? On the question of treating subjectivity objectively, see Schubert Ogden (1966) and Roger W. Sperry (1976). On *modern* science's commitment to the categories of objectivity alone, see Frederick Ferré (1976, chap. 1 and 2).

36. For suggestions as to what a post-modern science might be like, see Ferré (1976, chap. 5) and Stephen Toulmin (1982).

37. All references to Prigogine's writing in the text, unless otherwise indicated, are to *Order Out of Chaos: Man's New Dialogue with Nature,* by Ilya Prigogine and Isabelle Stengers (New York: Bantam Books, 1984).

38. See the discussion near the end of section 1.2, above.

39. Letter to the author, November 1, 1984.

40. Merrill, p. 240. A similar point, quoted by Emile Meyerson (1976, 353), was made by W. Wien: "Although the relationship between space and time as revealed by relativity theory is very important, one must always bear in mind . . . that it is not time itself which plays this role, but imaginary time."

41. Quoted by Čapek (1976, 366f.). However, as Čapek points out, Einstein was not much interested in this point, and vacillated on it. For example, the

first epigraph from him at the head of this essay suggests that time is a space-like dimension that can be fitted into a geometrization of reality.

42. The quotation is from James Jeans, *Man and the Universe* (1935).

43. This does not mean that *every* feature of the present universe is contingent. Rather, on the assumption that there has always been a plurality of events, in some state or other, it follows that there are some strictly eternal, or meta-physical, principles these events exemplify. Whitehead calls *Process and Reality* "an essay in cosmology." The task of cosmology is to describe the most general features of our cosmic epoch, while trying to distinguish the truly metaphysical aspects of reality from those aspects which apply only to our cosmic epoch. See Whitehead (1978, 287–88).

44. See Toulmin and Goodfield (1965, 247–65).

44a. That this fact is not widely understood is suggested by this statement by Karl Popper (1982, II, 5): "Since St. Augustine, at least, Christian theology has for the most part taught the doctrine of indeterminism; the great exceptions are Luther and Calvin." In reality, Luther and Calvin only differed from Augustine, Aquinas and others by being more explicit about their determinism. See the relevant chapters in Griffin (1976).

45. For an excellent account of the Hua-Yen school, with a critique of its idea of the symmetry of time from a Whiteheadian perspective, see Steve Odin, *Process Metaphysics and Hua-yen Buddhism: A Critical Study of Cumulative Penetration vs. Interpenetration* (Albany: State University of New York Press, 1982).

46. Capra (1975, 179) quotes this statement from D. T. Suzuki's explanation of the Zen perspective: "In this spiritual world there are no time divisions such as the past, present, and future; for they have contracted themselves into a single moment of the present. . . . the past and the future are both rolled up in this present moment of illumination." Suzuki goes on to say that "this present moment is not something standing still with all its contents, for it ceaselessly moves on." One cannot help wondering, if the past and future (as it actually will be) are already rolled up in the present, what it "moves on" *into*. Capra admits this difficulty; in fact, it is this difficulty that provides the rationale for his book, as he adds: "But modern physics may help" (179). But there is good reason to doubt that it can. For one thing, empirical evidence cannot resolve a logical contradiction. Also, the suggestion of process philosophy is that the aid can mainly go in the other direction. That is, it is time as known in our experience in its fullness, in which the future is in the present in a different mode than is the past, that can help us understand the reality of time in that aspect of the world physics studies.

47. See, for example, Ryusei Takeda, "Some Reflection on Yogācāra Philosophy in the Light of Whitehead's Metaphysics," unpublished essay, on file at the Center for Process Studies, 1325 N. College, Claremont, CA 91711.

48. I use the qualifier *so-called* because the term *precognition* imposes an unwarranted interpretation upon the experience. The term suggests that an event can be cognized (known) *before* it has occurred, which implies that the

event in some sense already exists to be known before it has appeared in the present. As G. J. Whitrow puts it: "unless we live in a 'block universe' in which everything in our future is pre-ordained so that we are mere automata completely unable to influence any of our future actions—a situation that seems to conflict with our actual experience—a future event is nothing but an unrealized possibility until it happens and therefore cannot give rise to genuine precognition" (Whitrow, 1980, 367f.). My only qualification would be that there *are* objective probabilities, and some so-called precognitive experiences might result from having an intuition of these. But this would not be genuine precognition. On the one hand, it would not be *cognition* of the future event as such, since cognition means *knowledge* and, so long as an event is merely probable, something *could* happen to prevent its occurrence; hence, that it is to occur cannot be *known* in advance. On the other hand, knowing a probability is knowing something that already exists and hence is not *pre*cognition.

49. Of course, many people reject all so-called paranormal occurrences out of hand. For them, there is no problem to be solved, since there are no precognitive experiences. Any "solution" that involves an appeal to psychokinesis and extrasensory perception is necessarily as absurd as the imagined problem it means to solve. However, my point here is only that those who *do* accept the reality of paranormal experiences in general can hold to the ultimate reality of temporality without arbitrarily denying the genuineness of that class of paranormal events that has led people to speak of "precognition." Those forms of paranormal events which do not suggest the causal influence of the future upon the present can adequately account for that one form which does, at first glance, seem to suggest this.

50. This example of the Holocaust was given by Wolf himself in a lecture for a meeting of the World Futures Society held at the University of Southern California on April 28, 1984. Although the idea of re-creating the past is not strongly featured in *Star Wave,* in this prepublication lecture, and in private conversation, he lifted up this idea as one of the chief implications of the book.

51. This kind of idea has been around for several decades. For example, G. N. Lewis (who coined the term *proton*) suggested that we could prevent light from having been emitted from a star a thousand years ago (Lewis, 1926, 25). Wolf is dependent primarily upon John Archibald Wheeler, who has defended the idea more recently. For a popular account, see Wheeler (1982, esp. 14–18).

52. Wolf *says* he does not affirm that the past and future are symmetrical, but that the past is open while the future is closed, i.e., strictly predestined (1984, vii, 326). However, in other passages, especially exhortatory ones, it is clear that he, like everyone else, in practice assumes that the future in part is still to be determined.

References

Augustine, Saint. 1912. *Confessions.* Cambridge: Harvard University Press.
Bergson, Henri. 1911. *Creative Evolution.* New York: Holt.
Bondi, H. 1952. "Relativity and Indeterminacy." *Nature* 169, April 19: 666.
Borges, Jorge Luis. 1967. "A New Refutation of Time." In *A Personal Anthology,* trans. Anthony Kerrigan, 44–64. New York: Grove Press.
Bridgman, Percy. 1955. *Reflections of a Physicist.* New York: Philosophical Library.
Brumbaugh, Robert S. 1984. *Unreality and Time.* Albany: State University of New York Press.
Čapek, Milič. 1961. *The Philosophical Impact of Contemporary Physics.* New York: Von Nostrand, Reinhold.
———. 1976. *The Concepts of Space and Time.* Boston Studies in the Philosophy of Science, Vol. 22. Dordrecht, Holland: Reidel.
———. 1983. "Time-Space rather than Space-Time," *Diogenes* No. 123 (July-September), 30–48.
Capra, Fritjof. 1975. *The Tao of Physics: An Exploration of the Parallels between Modern Physics and Eastern Mysticism.* Boulder: Shambala.
Carnap, Rudolph. 1963. "Intellectual Autobiography." In *The Philosophy of Rudolph Carnap,* ed. P. A. Schilpp, 3–43. La Salle, IL: Open Court.
Chisholm, Roderick. 1976. *Person and Object.* La Salle, IL: Open Court.
Cleugh, Mary F. 1937. *Time and Its Importance in Modern Thought.* London: Methuen.
Davies, P. C. W. 1976. *The Physics of Time Asymmetry.* Berkeley: University of California Press.
De Beauregard, Costa. 1966. "Time in Relativity Theory: Arguments for a Philosophy of Being." In Fraser (1966, 417–33).
De Broglie, Louis. 1949. "A General Survey of the Scientific Work of Albert Einstein." In *Albert Einstein: Philosopher-Scientist,* ed. P. A. Schilpp, 107–27. La Salle, IL: Open Court.
Denbigh, Kenneth G. 1975. *An Inventive Universe.* New York: Braziller.
———. 1981. *Three Concepts of Time.* New York: Springer-Verlag.
Dossey, Larry, M.D. 1982. *Space, Time and Medicine.* With a Foreword by Fritjof Capra. Boulder: Shambala.
Ebon, Martin. 1968. *Prophecy in Our Time.* New York: New American Library.
Eddington, Arthur. 1922. *The Theory of Relativity and Its Influence on Scientific Thought.* Oxford: Clarendon.
———. 1968. *The Nature of the Physical World.* Ann Arbor: University of Michigan Press.
Einstein, Albert. 1950. *The Meaning of Relativity.* 3d ed. Princeton: Princeton University Press.
———. 1954a. *Ideas and Opinions.* Ed. Carl Seelig. New York: Crown.
———. 1954b. *Relativity: The Special and the General Theory,* Trans. R. W. Lawson. London: Methuen.

————. 1976. "Comment on Meyerson's 'La Deduction Relativiste.' " In Čapek (1976, 363–67).

Eisenbud, Jule. 1956. "Psi and the Problem of the Disconnections in Science." *Journal of the American Society for Psychical Research* 50:3–26.

Feigl, Herbert. 1960. "Mind-Body, *Not* a Pseudoproblem." In *Dimensions of Mind,* ed. Sydney Hook, 24–36. New York: New York University Press.

Ferré, Frederick. 1972. "Grünbaum on Temporal Becoming: A Critique." *International Philosophical Quarterly* 12: 426–45.

————.1976. *Shaping the Future: Resources for the Post-Modern World.* New York: Harper & Row.

Feynman, Richard. 1965. *The Character of Physical Law.* Cambridge: M.I.T. Press.

Frank, Phillip. 1976. "Is the Future Already Here?" In Čapek (1976, 387–95).

Fraser, J. T. 1966. *The Voices of Time.* New York: Braziller.

————. *Of Time, Passion and Knowledge.* New York: Braziller.

Freeman, Eugene, and Sellars, Wilfrid, eds. 1971. *Basic Issues in the Philosophy of Time.* La Salle, IL: Open Court.

Gale, Richard M. 1968. *The Language of Time.* New York: Humanities Press.

Gold, Thomas, ed. 1967. *The Nature of Time.* Ithaca, NY: Cornell University Press.

Gold, Thomas. 1977. "Relativity and Time." In *The Encyclopedia of Ignorance,* ed. R. Duncan and M. Weston Smith. New York: Pergamon.

Griffin, David Ray. 1976. *God, Power, and Evil: A Process Theodicy.* Philadelphia: Westminster.

————. 1981. "Creation Out of Chaos and the Problem of Evil." In *Encountering Evil: Live Options in Theodicy,* ed. Stephen T. Davis, 101–19. Atlanta: John Knox.

Grünbaum, Adolf. 1963. *Philosophical Problems of Space and Time.* New York: Knopf.

————. 1967a. "The Anisotropy of Time." In Gold (1967, 149–86).

————. 1967b. *Modern Science and Zeno's Paradoxes.* Middletown: Wesleyan University Press.

Hoffman, Banesh (with Helen Dukas). 1972. *Albert Einstein: Creator and Rebel.* New York: Viking Press.

Lawrence, Nathaniel. 1971. "Time Represented as Space." In Freeman and Sellars (1971, 123–32).

Lewis, G. N. 1926. "The Nature of Light." *Proceedings of the National Academy of Sciences* 12:22–29.

————. 1930. "The Symmetry of Time in Physics." *Science* 71:569–77.

Matson, Floyd W. 1966. *The Broken Image: Man, Science and Society.* Garden City: Doubleday.

Mehlberg, Henry. 1980. *Time, Causality, and the Quantum Theory,* vol. 1. Ed. Robert S. Cohen. Dordrecht, Holland: Reidel.

Merrill, A. A. 1922. "The *t* of Physics," *Journal of Philosophy* 19 (9), April 27, 238–41.

Meyerson, Emile. 1930. *Identity and Reality.* Trans. Kate Loewenberg. London: Allen & Unwin.

———. 1976. "Various Interpretations of Relativistic Time." In Čapek (1976, 353–62).

Monod, Jacques. 1972. *Chance and Necessity.* New York: Vintage Books.

Needham, Joseph. 1943. *Time: The Refreshing River.* New York: Macmillan.

Ogden, Schubert. 1966. "Theology and Objectivity." In *The Reality of God and Other Essays,* 71–98. New York: Harper & Row.

Popper, Karl R. 1977. (with John C. Eccles), *The Self and Its Brain* (Springer).

———. 1982. *Postscript to the Logic of Scientific Discovery,* ed. W. W. Bartley, III (Rowman and Littlefield), Vol. II: *The Open Universe: An Argument for Indeterminism;* Vol. III: *Quantum Theory and the Schism in Physics.*

Prigogine, Ilya. 1973. "Time, Irreversibility and Structure." In *The Physicist's Conception of Nature,* ed. Jagdish Mehra, 561–93. Dordrecht, Holland: Reidel.

———. 1980. *From Being to Becoming.* San Francisco: W. H. Freeman.

Prigogine, Ilya, and Stengers, Isabelle. 1984. *Order Out of Chaos: Man's New Dialogue with Nature.* New York: Bantam Books.

Quine, Willard. 1960. *Word and Object.* Cambridge: M.I.T. Press.

Reichenbach, Hans. 1956. *The Direction of Time.* Berkeley: University of California Press.

———. 1958. *The Philosophy of Space and Time.* New York: Dover.

Russell, Bertrand. 1917. *Mysticism and Logic.* London: Allen & Unwin.

———. 1921. *Our Knowledge of the External World.* London: Allen and Unwin.

Sipfle, David A. 1971. "On the Intelligibility of the Epochal Theory of Time." In Freeman and Sellars (1971, 181–94).

Sperry, Roger W. 1976. "Mental Phenomena as Causal Determinants in Brain Function." In *Consciousness and the Brain: A Scientific and Philosophical Inquiry,* ed. G. Globus, G. Maxwell, and I. Savodnik, 163–77. New York: Plenum Press.

Steuwer, Roger. 1975. "G. N. Lewis on Detailed Balancing, the Symmetry of Time, and the Nature of Light." In *Historical Studies in the Physical Sciences,* vol. 6, ed. Russell McCormmach, 469–511. Princeton: Princeton University Press.

Suppes, Patrick, ed. 1973. *Space, Time and Geometry.* Dordrecht, Holland: Reidel.

Tanagras, Angelos. 1967. *Psychophysical Elements in Parapsychological Traditions.* Parapsychology Foundation.

Toulmin, Stephen. 1982. *The Return to Cosmology: Postmodern Science and the Theology of Nature.* Berkeley: University of California Press.

Toulmin, Stephen, and Goodfield, June. 1965. *The Discovery of Time.* New York: Harper & Row.

Weyl, Herman. 1949. *Philosophy of Mathematics and Natural Science.* Princeton: Princeton University Press.

Wheeler, John Archibald. 1982. "Bohr, Einstein, and the Strange Lesson of the Quantum." In *Mind in Nature: Nobel Conference XVII,* ed. Richard Q. Elvee, 1–30. New York: Harper & Row.

Whitehead, Alfred North. 1926. *Science and the Modern World.* 2d ed. New York: Macmillan.

———. 1927. "Time." *Proceedings of the Sixth International Congress of Philosophy,* 59–64. New York and London: Longmans, Green.

———. 1933. *Adventures of Ideas.* New York: Macmillan.

———. 1947. *Essays in Science and Philosophy.* New York: Philosophical Library.

———. 1958. *The Function of Reason.* Boston: Beacon Press.

———. 1959. *Symbolism: Its Meaning and Effect.* Capricorn.

———. 1964. *The Concept of Nature.* Cambridge: Cambridge University Press.

———. 1966. *Modes of Thought.* New York: Free Press.

———. 1978. *Process and Reality.* Corrected edition, ed. David Ray Griffin and Donald W. Sherburne. New York: Free Press.

Whitrow, G. J. 1980. *The Natural Philosophy of Time,* 2d ed. Oxford: Clarendon.

Wilber, Ken, ed. 1982. *The Holographic Paradigm and Other Paradoxes.* Boulder: Shambala.

Wolf, Fred. 1981. *Taking the Quantum Leap: The New Physics for Non-Scientists.* San Francisco: Harper & Row.

———. 1984. *Star Wave: Mind, Consciousness, and Quantum Physics.* New York: Macmillan.

Zukav, Gary. 1979. *The Dancing Wu Li Masters: An Overview of the New Physics.* London: Rider/Hutchinson.

Zwart, P. J. 1973. "The Flow of Time." In Suppes (1973, 131–56).

———. 1976. *About Time, A Philosophical Inquiry into the Origin and Nature of Time.* Amsterdam: North-Holland.

I   Historical Backgrounds

2. Andrew G. Bjelland

EVOLUTIONARY EPISTEMOLOGY, DURATIONAL
METAPHYSICS, AND THEORETICAL PHYSICS:
ČAPEK AND THE BERGSONIAN TRADITION

In the course of the century since the death of Charles Darwin, assessments of the philosophical relevance of evolutionary theory have varied widely. John Dewey, writing in 1909, heralded a revolution in philosophic thought—a revolution sparked by Darwin's contributions to biological theory:

> The conceptions that had reigned in the philosophy of nature and knowledge for two thousand years . . . rested on the assumption of the superiority of the fixed and final; they rested upon treating change and origin as signs of defect and unreality. In laying hands upon the sacred ark of absolute permanency, in treating the forms that had been regarded as types of fixity and perfection as originating and passing away, the *Origin of Species* introduced a mode of thinking that in the end was bound to transform the logic of knowledge, and hence the treatment of morals, politics, and religion.[1]

At the time of Dewey's comment, the impact of evolutionary theory was evident, to a greater or lesser extent, in the epistemologies of Spencer, Helmholtz, Avenarius, Mach, Ribot, Fouillée, Poincaré, Bergson, Peirce, James, and of Dewey himself. These diverse thinkers, whatever their differences, united in their endorsement of a common thesis: the structure of human reason, far from being suprahistorical and immutable, is the product of a long evolutionary process, in the course of which the human mind gradually adjusted itself to that sector of reality which possesses biological significance for the human species.[2] The variant forms of the evolutionary or biological theory of knowledge expressed the conviction that human cognitive activities, as functions of the total psychophysical organism, "cannot be exempt from the evolutionary process of which the organism itself is a result."[3]

This evolutionary trend was eclipsed by the rise of logical positivism in the 1920s and by subsequent developments in the logical-empiricist wing of what is loosely termed "analytic philosophy." In his *Tractatus logico-philosophicus* of 1921, Ludwig Wittgenstein expressed this contrasting view: the relation of evolutionary theory to philosophy is wholly external.

4.1122 Darwin's theory has no more to do with philosophy than any other hypothesis in natural science.

4.113 Philosophy settles controversies about the limits of natural science.

4.114 It must set limits to what can be thought; and, in doing so, to what cannot be thought.[4]

Philosophy, as setting the limits to what can be thought and expressed, may influence biology and the other natural sciences, but there seems to be no prospect that biology or any other natural science might influence the development of philosophy. Although Wittgenstein cautions his readers that "What expresses *itself* in language," namely, logical form, "we cannot express by means of language" (4.121), he nowhere in the *Tractatus* hints that logical form itself may be a product of biological and sociopsychological evolution. Natural languages are historical products; hence, they are conventional. The logical form that expresses itself in natural languages, however, as "the logical form of reality" (4.121)—that is, as the logical form of all possible worlds or of all possible universes of discourse—seems to transcend time and history.

The nonevolutionary approach articulated in the *Tractatus* came to dominate epistemological considerations for the remainder of the first half of the present century. The logical-empiricist program, in its opposition to psychologism, stressed the autonomy of logic, and, in advocating the "linguistic turn," all but eliminated significant references to experience. Elucidations of the logical structure of scientific explanation proceeded in isolation from any in-depth phenomenology of perception and experience. Although expressly "antimetaphysical" in intent and "conventionalist" in its main orientation, this neopositivist program in fact embodied a metaphysical and fundamentally rationalistic insistence on the applicability of a preferred logicomathematical framework to *any possible world.* This rationalistic facet of the empiricist program is evident in the following passage from A. J. Ayer's neopositivist manifesto, *Language, Truth, and Logic* (1946):

We saw that the reason why they [analytic propositions] cannot be confuted by experience is that they do not make any assertion about the empirical world. They simply record our determination to use words in a certain fashion. We cannot deny them without infringing the conventions which are presupposed by our very denial, and so falling into self-contradiction. And this is the sole ground of their necessity. As Wittgenstein put it, our justification for holding that the world could not conceivably disobey the laws of logic is simply that *we could not say of an unlogical world how it would look.* And just as the validity of an analytic proposition is *independent of the nature of the external world;* so it is *independent of the nature of our minds.*[5]

Few passages better illustrate what may be termed the "formalistic bias": the twofold view that some privileged logistic establishes not only the limits of our developed languages, but of all possible languages, and that language, in its turn, sets not only the limits of our *de facto* world, but of all possible worlds. As a consequence of this bias, what begins as linguistic conventionalism terminates in the expression of an epistemological and metaphysical Absolute—the Absolute "represented by analytical propositions, that is by logico-mathematical tautologies, applicable to any kind of world."[6] This bias, although now seldom so baldly endorsed as in the above passage, remains influential. It continues to inspire the belief that evolutionary and biological approaches to epistemological issues inevitably will be vitiated by the genetic and psychologistic fallacies. As a consequence, the dominant lines of twentieth-century Anglo-American philosophy until quite recently have remained inimical to evolutionary epistemologies.[7]

Developments in the past fifteen to twenty years indicate something of a change in the philosophical climate. In his *Human Understanding* (1972), Stephen Toulmin defends the thesis that "the proud self-sufficiency of philosophical epistemology will turn out to be, in actual fact, a mark of its sheer irrelevance."[8] Then, in a move that has parallels in the writings of Karl R. Popper and Donald T. Campbell, Toulmin advocates the view that the natural-selection paradigm of Darwinian theory can be generalized into a variation-and-selective-retention model for epistemological inquiry.[9] Campbell himself indicates that a generalization such as that recommended by Toulmin, even though it takes "cognizance of and [is] compatible with man's status as a product of biological and social evolution," need not issue in the genetic fallacy. Within the context of this evolutionary approach it is not claimed that knowledge at every level is merely an instrument in the service of survival interests; rather, the theory urges "that evolution—even in its biological aspects—is a knowledge process."[10]

Milič Čapek, in numerous articles and in two books, *The Philosophical Impact of Contemporary Physics* (1961)[11] and *Bergson and Modern Physics* (1971), has carefully articulated the character of the biological and evolutionary approach to epistemological concerns and has artfully delineated the relevance of this epistemological approach for a reassessment of fundamental issues in metaphysics and physics. In particular, Čapek, more than any other author, has explicated the interrelations of Henri Bergson's biological epistemology, theory of durational succession, protomentalist metaphysics, and early anticipations of subsequent developments in twentieth-century physics. Čapek not only provides a critical restatement of the interconnections linking these aspects of Bergson's thought, but, drawing on his own extensive knowledge of twentieth-century developments in physics and philosophy, offers additional support for, and elaborations of, Bergson's philosophy.

The first section of this paper reviews Čapek's analysis of Herbert Spencer's classical formulation of the biological theory of knowledge and his statement of Bergson's critical amendment of Spencer's theory. The second section displays Čapek's interpretation of the Bergsonian theory of durational succession and of its foundational role with respect to Bergson's pluralistic, protomentalist metaphysics. In my concluding comments, I address—from a perspective I believe to be continuous with Čapek's extension of the Bergsonian tradition—a number of issues concerning the interrelations of evolutionary epistemology, durational metaphysics, and theoretical physics.

2.1. Spencer's Biological Theory of Knowledge and Bergson's Amendment of Spencer's Theory

The main points of Čapek's analysis of Spencer's theory (cf. BMP 4–14) may be organized with reference to the claims I have italicized in the following passage from Spencer—a passage that Čapek cites as embodying the key features of Spencer's position:

[1] *Life in its simplest form is the correspondence of certain inner physicochemical actions with certain outer physico-chemical actions, [and] each advance to a higher form of Life consists in a better preservation of this correspondence by the establishment of other correspondences.* . . . Life is definable as the continuous adjustment of internal relations to external relations. And when we so define it, we discover that physical and psychical life are equally comprehended by that definition. . . . [2] *Intelligence . . . arises when the external relations to which the internal ones are adjusted, begin to be numerous, complex and remote in time or space.* [4] *Every advance in Intelligence essentially consists in the establishment of more varied, more complete, or more involved adjustments. And even the highest generalizations of science consist of mental relations of coexistence and sequence, so coordinated as exactly to tally with certain relations of coexistence and sequence that occur externally.* . . . . [When we pose questions of a theoretical nature to the chemist or mathematician,] we . . . find that special or general relations of coexistence and sequence between properties, motions, spaces, etc., are all they can teach us. [3] *And lastly, let it be noted that what we call truth, guiding us to successful action and the consequent maintenance of life, is simply the accurate correspondence of subjective*

to objective relations; while error, leading to failure and therefore towards death, is the absence of such accurate correspondence.[12]

The mechanistic character of Spencer's theory is evident. Bracketed numbers in the above passage indicate the chief features of Spencer's epistemology and correspond with the order in which these specific features are dealt with below.

1. *The Continuity of Life and Mechanistic Processes.* For Spencer the phenomena of life, including human cognitive functions, are in no sense independent of underlying mechanistic processes. Mental life and the material environment are both expressions of a more fundamental reality—which in theory is to be designated solely as "the Unknowable" but which in practice is identified with the "objective order" depicted in the corpuscular-kinetic picture of nature.[13] Spencer's biological theory of knowledge is expressed in terms wholly consonant with his basic commitment to mechanistic monism.

2. *The Continuity of Infrahuman and Human Intelligence.* Intelligence, as an internal organic adjustment to external environmental relations, is continuous with other vital functions and is distinguished from simpler adjustments solely in terms of the number, the complexity, and the spatiotemporal range of the external factors with which it is correlated. In the context of Spencer's associationist psychology, there is no boundary separating sensory perception, and its extensions in imagination and memory, from intelligence, nor is there a definitive demarcation between infrahuman and human intelligence (cf. BMP 6–7).

3. *The Survival Value of Intelligence and the Biological Utility of Truth.* Although *truth* and *error* are employed primarily with reference to human affairs, they may be analogically extended with respect to the activities of other species. For Spencer the presence or absence of a biologically useful adjustment to the environment constitutes the behavioral definitions of "truth" and "error." Thus knowledge and intelligence, when construed in Spencer's broad sense, are primarily directed to practical and biological ends; they are instruments in the struggle for life, which, in securing the individual's survival, perpetuate the life of the species (cf. BMP 6–7, 36).

4. *Scientific Intelligence: The Culmination and Completion of Evolution.* The fourth claim implies that the evolutionary process presently *culminates* in that correlation of subjective

to objective relations which *is* human intelligence. When set within the larger context of Spencer's system and when viewed against the broad backdrop of later-nineteenth-century positivistic thought, however, this claim further implies that, with the development of scientific intelligence and the corpuscular-kinetic world view of classical physics, the evolutionary process both *terminates* and—in principle, if not in detail—*is completed.* Developed scientific intelligence provides the world-machine with its precise reflective image; thus with the advent of scientific consciousness, evolutionary process becomes, so to speak, self-aware.

In Spencer's view, as in that of the vast majority of late-nineteenth-century positivists, the rational structures exhibited in Euclidean geometry and Newtonian physics represent the complete adjustment of subjective to objective relations. Indeed, Spencer stands in fundamental agreement with the Kantian view that the foundations of the Newtonian-Euclidean world-picture are immune to revision: they can never be contradicted by future experience. In Spencer's evolutionist vision, however, this impossibility of future confutation is not grounded in any supposedly permanent and *a priori* structure of the human mind. The forms and categories do not possess validity because they are, in any Kantian sense, prior to or independent of experience (cf. FP 159*n*). For Spencer the forms and categories are abstract schemata that, as derived from the concrete particularity of experience, have been progressively stabilized in the evolving neurological structures of the human organism and in the patterns of language and social practice. This stabilizing process occurred "throughout the entire evolution of intelligence"; consequently, such foundational schemata "have been rendered organic by immense accumulation of experience, received partly by the individual, but mainly by all ancestral individuals whose nervous system he inherits" (FP 147, 159). As products of evolution and as deeply entrenched in language, these structures possess a relatively *a priori* status for any given individual; thus they confront individual consciousness with all the force of fundamental intuitions. Spencer in his distinctive manner thus endorses the "transcendental" status of the Newtonian-Euclidean framework of classical science.[14]

Spencer's position may be seen as an attempted mediation of the rationalist-empiricist polarization; his attempted mediation, however, remains weighted on the side of empiricism. The Lockean-Humean category of *individual* experience is replaced by the Spencerian notion of a cumulative, species-wide experience. This broadening of the concept of experience, however, issues in a distinctive approach to the synthetic *a priori.* Within the context of Spencer's associationism, the subjectivist "inconceivability-of-the-opposite" test and its counterpart, the "intense-feeling-of-evidence" test, may be presumed to provide efficacious criteria for identifying those truths which are *both* descriptive of the actual, objective physical universe *and* immune to revision. Successful beliefs are adjustments to the environing world that possess survival value. Unsuccessful beliefs, as maladjustments, are devoid of utility and are progressively eliminated. The feelings that accompany successful beliefs also serve survival interests. The intense feelings of evidence accompanying the foundational beliefs of science and the intense feelings of absurdity following in the wake of attempts to deny them are thus subjective correlates of the objective truth or survival value of those beliefs.[15]

For Spencer, it is virtually inconceivable that the general beliefs that ground the Newtonian-Euclidean world view—beliefs that have survived innumerable tests throughout the evolutionary development of the species—could ever be subject to revision. In their practical immunity to revision, such beliefs display the chief characteristic of necessary truths. Spencer concludes that those logicomathematical and scientific propositions that present themselves to consciousness as immune to revision are possessed of a biopsychological necessity that is ultimately grounded in external, objective necessity. This objective necessity constitutes not the Logos of all possible experience or of all possible worlds, but rather the Logos of that one actual world in which we, as biological entities, live, and act, and sustain our being.[16]

The stage is now set for a proper understanding of Bergson's critical reformulation of the biological theory of knowledge. Bergson provides several accounts of Spencer's role in the development of his own distinctive philosophy.[17] Of these, none is more generally illuminating than that communicated to Charles Du Bos in 1922. In his conversation with Du Bos, Bergson noted that during the 1880s two opposed factions dominated the philosophical life of the university. Those who held that Kant had definitively formulated the major philosophical issues constituted the far larger group. Bergson's sympathies, however, linked him with the minority, who supported the evolutionism of Spencer. Bergson was initially attracted to Spencer's philosophy by the concrete character of the British thinker's efforts to delineate the Logos of the actual world. For the young Bergson, Spencer's "desire always to bring the mind back upon the terrain of facts" (OE 1541) issued in an empiricism that stood in marked opposition to the Kantian edifice and its rationalistic attempt to establish the *a priori* limits of *all possible experience.*

By 1881 Spencer's thought commanded Bergson's unreserved allegiance. Indeed, he initially envisioned his own doctoral dissertation as comprising a continuation, consolidation, and completion of Spencer's account of the foundations of science.

During 1883–1884 Bergson critically reexamined Spencer's account, in his *First Principles,* of the foundational notions of mathematics and mechanics. Analysis of the notion of time utilized in classical mechanics—of that time which, like space, is a homogeneous medium containing juxtaposed parts and which consequently is amenable to precise mathematical description—led Bergson to the vague awareness that any attempt to identify experienced temporality with scientific time must issue in insurmountable difficulties. Further reflections on the notion of that time which grounds scientific objectifications, and on Zeno's paradoxes, led him to conclude that scientific time *does not endure in any sense:* "that nothing would have changed in our scientific knowledge of things if the totality of the real had been unfurled all at a stroke in an instant, and that positive science essentially consists in the elimination of duration."[18] With this realization, Bergson abandoned his projected completion of Spencer's account of the foundations of science.

Spencer's philosophy of science does provide a clear illustration of the manner in which positivism eliminates duration. For Spencer, the principle of the persistence of Force (the principle of the conservation of energy) is the most fundamental law of physics "which, as being the basis of science, cannot be established by science" (FP 175). The foundational role Spencer assigns to this principle grounds his adherence to the strictly deterministic, Laplacean conception of causality. For Spencer, as for Laplace, *"causa aequat effectum"* in the sense that the cause-effect relation is one of logical equivalence: from any given "present" state of the universe, each and every "future" and "past" state is in principle derivable. Within the context of the principle of the persistence of Force, the formula *"causa aequat effectum"* finds "expression in the quantitative equality of successive energetic equivalents" (BMP 12). Spencer, like most nineteenth-century positivists, could believe that the mysterious link of necessity, mutually binding cause and effect—the link that had eluded Hume—had been discovered: "No cause can be without its effect because no particular quantity of energy can disappear without being transformed into its equivalent; and no effect can be without its cause because no quantity of energy . . . can arise out of nothing, but only from the transformation of a previous energetic equivalent" (BMP 13).

In Spencer's view these considerations imply that the law of continuous redistribution of matter and motion must find expression both in processes of evolution and in processes of devolution. Moreover, the validity of the law of the persistence of Force may be extrapolated in both directions of time, thereby securing the temporal uniformity of nature. Finally, the logic of mutual implication, if it were to remain strictly in effect, would entail the conclusion that there is "no intrinsic difference between the two directions of time [other than] the difference comparable to the conventional difference between a plus and a minus sign in analytical geometry" (PICP 124–25). The asymmetry of the relation of past to present—the qualitative asymmetry of experienced durational succession—would be banished to the realm of "merely subjective" phenomena. If the intrinsic logic of Spencer's foundational concepts were to be rigorously developed, his "evolving" cosmos would be indistinguishable from the static and deterministic universe of Laplace.[19]

Bergson, in opposition to the main lines of Spencer's thought, became increasingly convinced that lived-time, the time of development, growth, and emergent novelty, or the duration "which plays the leading part in any philosophy of evolution, eludes mathematical treatment" (CM 12 = OE 1254). Lived-duration, the essence of which is to flow, does not admit of the superposition "of one part on another with measurement in view" (CM 12 = OE 1254). Whereas duration is truly effectual in the sense that it retards and "hinders everything from being given at once," in Spencer's mechanistic evolutionism "time served no purpose, did nothing" (CM 93 = OE 1333). Spencer's "empirical evolutionism," far from evidencing fidelity to the facts of the evolving, was thoroughly vitiated by its acceptance of evolved symbol—the static, homogeneous "time" of mechanics—as adequately representing evolving reality. Consequently, *the usual device of the Spencerian method consists in reconstructing evolution with fragments of the evolved"* (CE 396 = OE 803).[20] That fidelity to facts which had attracted Bergson's interest to Spencer's thought was at best partial. Within the context of Spencer's sensistic, associationist psychology, foundational "facts" were all too tinged by what Whitehead later would term "the fallacy of misplaced concreteness"—Spencer's presentation of "facts" neglected the degree of abstraction involved when experience is viewed as exemplifying the formal categories of thought.

Bergson continued to endorse Spencer's "view that the present form of human intellect is a result of the gradual evolutionary adaptation of the human psychophysical organism to the order of nature" (BMP 30); however, he came to reject Spencer's contention that with the advent of the Newtonian-Euclidean (or, better, Laplacean-Euclidean) form of intellect, the adaptational correspondence of subjective to objective relations was in principle complete. Experienced duration eludes reduction to the spatialized time of classical mechanics. Reflection on this fact led Bergson to conclude that a consistent and experientially grounded biological theory of knowledge issues in the following thesis: "The Laplacean-Euclidian form of intellect *does not adequately represent nature in its entirety, but merely the sector of it which is of vital importance for the human organism*" (BMP 30).

Bergson's universe, like that later depicted in Reichenbach's *Atom and Cosmos*,[21] is a universe of dynamically diversified strata, not all of which are amenable to the corpuscular-kinetic interpretations of classical physics. Čapek notes: "What is 'intelligible' in the classical sense is what Reichenbach appropriately called 'the world of middle dimensions' located between microcosmos and 'megacosmos', i.e. between the zone of atomic events and the universe as a whole: *if* there is a whole!" (BMP 31). Our Newtonian-Euclidean logic of solid bodies and our object-oriented language represent the human organism's evolutionary adjustment *not* to the totality of the real, *but rather* to those properties of the real which are biologically significant. As a consequence, Spencer's theory must be amended as follows: *"The limited applicability of the classical . . . modes of thought is due to the fact that they themselves are products of evolutionary adjustment to a limited segment of reality: consequently, when by the process of extrapolation we try to apply them outside of the zone to which they are adjusted, their inadequacy becomes obvious—the more so, the further beyond the limits that they are applied"* (BMP 31).

Cognitive functions, whether perceptual or intellectual, more effectively serve the organism if they are selective; hence, the limited applicability of our habitual categories is grounded in the immanent teleology of the human organism (cf. BMP 33). Organisms, including *Homo sapiens,* are not disinterested spectators; they are agents actively exploring the environment with a view to satisfying their needs and to thereby sustaining their activity. From *Matter and Memory* (1896) onward, Bergson rejects the postulate, common to both empiricist and rationalist epistemologies, that *"perception has a wholly speculative interest"* [22] and promotes the counterthesis that human cognition, even in its presumably most passive form, sensory perception, is essentially active and selective.

Cognitive functions, because they are selective, simplify, but do not falsify, the structure of the world. Our cognitive simplifications have their objective grounding in the immanent teleology and economy linking organism and environment. These simplifications, inasmuch as they serve the needs and interests of action, cannot be grounded in mere subjective illusion (cf. CE xxi = OE 491, and BMP 282). Our categories are well-founded simplifications and approximations of the objective environment; they are dynamically and intentionally relational, but not relativistic in any radically subjectivist, idiosyncratic sense. They enjoy objective, but limited, applicability. Error results, however, if their use is extended beyond the limited range of their applicability.

Bergson's biological theory of knowledge in its first, primarily critical and negative stage of development, looks to the structures of *the evolved* and cautions us against the overextension of our habitual cognitive categories. This aspect of the theory constitutes the core of Bergson's often noted and frequently misrepresented "anti-intellectualism." This "anti-intellectualism"—the claims concerning the inadequacy of the Newtonian-Euclidean intellect and the necessity of intuitive knowledge—cannot in fairness be equated with a thoroughgoing irrationalism. Bergson does not attack human reason as such, nor does he issue an unrestricted condemnation of even the Newtonian-Euclidean form of intellect and its stabilizing categories; rather, he notes the limited applicability of our habitual modes of thought, insists that their true significance cannot be divorced from the immanent teleology linking organism and environment, and criticizes the errors that abound when these categories are employed outside the limited range of their applicability.

A central insight grounds Bergson's recognition of the limited applicability of Newtonian-Euclidean categories and his consequent rejection of both the Kantian and Spencerian scientific epistemologies: lived-duration eludes reduction to the abstract schema of objectifying time. The negative and critical phase of Bergson's thought has as its positive counterpart his phenomenology and metaphysics of lived-duration. On its constructive side, Bergson's theory looks to the dynamic processes of *the evolving* and issues in an authentically dynamic, evolutionary metaphysics of durational experience and agency.

2.2. Durational Succession and Bergsonian Protomentalism

On its positive side, the Bergsonian system progressively displays the consequences, for philosophy and science, flowing from a single, central thesis: durational succession—whether psychological or physical—is a dynamic, asymmetrical, and internal relation. This thesis, indeed, constitutes the very heart of the Bergsonian system and lends that system its distinctive, revolutionary character. The present section (1) analyzes Bergson's theory of durational succession; (2) displays the continuity of that theory with Bergson's protomentalist metaphysics; (3) contrasts Bergson's theory with the theories of temporal succession that dominated thought during the classical period of Western philosophical and scientific development; and (4) links Bergson's theory of durational succession with his theory of causality and protomental agency—a theory of causation that endorses the objective indetermination of the future.[23]

The thesis that durational succession is a dynamic, asymmetrical, and internal relation undergirds Bergson's metaphysical research program, which, like that of Whitehead, is best viewed as a version of panpsychistic or protomentalistic pluralism.[24] The following passage from *Duration and Simultaneity* (1922) provides a strong statement of the link between Bergson's theory of durational succession and his protomentalist metaphysics:

What we wish to establish is that we cannot speak of a reality that endures without inserting consciousness into it. . . . We may perhaps feel averse to the use of the word "consciousness" if an anthropomorphic sense is attached to it. But to imagine a thing that endures, there is no need to take one's own memory and transport it, even attenuated, into the interior of the thing. . . . It is the opposite course we must follow. We shall have to consider a moment in the unfolding of the universe, that is, a snapshot that exists independently of any consciousness, then we shall try conjointly to summon another moment brought as close as possible to the first, and thus have a minimum of time enter into the world without allowing the faintest glimmer of memory to go with it. We shall see that this is impossible. Without an elementary memory that connects the two moments, there will be only one or the other, consequently a single instant, no before and after, no succession, no time. We can bestow upon this memory just what is needed to make the connection; it will be, if we like, this very connection, a mere continuing of the before into the immediate after with a perpetually renewed forgetfulness of what is not the immediately prior moment. We shall nonetheless have introduced memory. To tell the truth, it is impossible to distinguish between the duration, however short it may be, that separates two instants and a memory that connects them, because duration is essentially a continuation of what no longer exists into what does exist. This is also any conceived time, because we cannot conceive a time without imagining it as perceived and lived. Duration therefore implies consciousness; and we place consciousness at the heart of things for the very reason that we credit them with a time that endures.[25]

Following Milič Čapek's suggestion (cf. BMP 303), we may interpret the foregoing text as a thought experiment designed to test the applicability of the notion of matter displayed in Leibniz's formula: *"Omne enim corpus est mens momentanea sive carens recordatione"* ("Every body is an instantaneous mind or a mind lacking recollection").[26] This thought experiment terminates in the realization that even a perpetually perishing material process—a process "which dies and is born again endlessly" (CE 220 = OE 665) after the manner of "a perpetually renewed forgetfulness"—must have a minimal temporal thickness. Not even physical events can be characterized, in the strict sense, as "instantaneous"—as isolated, self-contained, and as externally related to their past. Instantaneity is a limit concept. These reflections contain, in germ, the doctrine that durational succession, as a relation, must be dynamic, asymmetrical, and internal. We may now flesh out the Bergsonian theory of succession in detail.

Durational succession is a dynamic relation.

Durational succession, as a "becoming of continuity," is a dynamic relation linking past and present. It is the dynamic "continuation of what no longer exists" as the concrete immanence of self-actualizing process, "into what does exist" *in the strictest sense* as the concrete present of immanent actualization. The terms of this relation—past and present, before and after—arise only within the context of becoming. The dynamics of that context exclude the coexistence of the relata—and thereby exclude the reduction of durational succession to spatial juxtaposition.

Durational succession is an asymmetrical relation.

The succession of one physical event by the next involves the emergence of a novel present that, no matter how conformally continuous it is with its causal past, is not identical with its past. The novelty of the present is impossible apart from the persistence of the past as its qualitatively contrasting ground. Similarly, the pastness of the previous event is unthinkable without the novelty of the present. The *asymmetry* or *irreversibility* of the relation is a direct consequence of this dynamic, qualitative ordering of its successive terms. This asymmetry, moreover, entails the irreducibility of the causal order to the static order of logical equivalence and identity—to that static order implicit to the dictum: *"Causa aequat effectum."* The advance of present over past, if highly conformal, may approach the identity of an atomic, durationless instant as an ideal limit, but, precisely as durational advance, cannot coincide with that limit. Achieved coincidence would entail the collapse of succession.

Durational succession is an internal relation.

The sheer advance of present over past involves the ascription of some minimal novel agency to the present event. Since "it is impossible to distinguish between the duration, however short it may be, that separates two instants and a memory that connects them," this emergent novel agency, internal to the present event, must be mnemic. Durational succession is intelligible only in terms of the qualitative differentiation of present from past, which differentiation itself depends on the mnemic survival of the past in the present.

This mnemic survival of the past within the present event cannot be an *external* relation, for, as Čapek notes, "both terms, despite their succession and despite their difference—*or rather because of it*—are not separated. . . . [The] qualitative difference 'separating' two successive moments at the same time *joins* them" (BMP 130). The relation, in terms of the very meanings of "past" and "present," is one of mutual internality: the past as *past* and the present as *present* are coherently meaningful solely in virtue of a mutuality of qualitative contrast, which is also the basis of their unique continuity precisely as successive.

In the Bergsonian theory of durational succession, the relation of past to present displays Whiteheadian "coherence": past and present, as fundamental ideas in terms of which the theory is developed, "presuppose each other so that in isolation they are meaningless."[27] This coherence does not require that "past" and "present" be definable in terms of one another; rather, what is indefinable in the one notion cannot be abstracted from its relevance to the other. Past event and present event are related *because of,* not in spite of, the novelty that is the basis for their differentiation.

The mutual internality of the meanings "past" and "present" is not inconsistent with the dynamic asymmetry of the concrete relation linking present event to past event. This nexus occurs in virtue of the prehensive character of the emergent present event. The determinate past, as object, is not identical with the present as act—with the present in its character as mnemic, prehensive synthesis. Consequently, the past event's survival in the present event is abstractive and intentional rather than concrete and fully actual. With respect to physical events, the past, as causally efficacious, is immanent to the present; yet the present, as novel, mnemic synthesis, is asymmetrically related to the past.

The theory that durational succession is a dynamic, asymmetrical, and internal relation bears the experiential warrant of applicability. The chief competing theories bear no such warrant. The revolutionary character of Bergson's thought is evident when his theory of succession is contrasted with those alternative theories—often unvoiced—which dominated reflection throughout the classical modern period of Western philosophical and scientific development.

The attempt to interpret succession as an external relation—the mainline tendency of atomistic empiricisms, whether logical or psychological—is deeply rooted in Hume's central logical doctrine: "Whatever is distinct, is distinguishable; and whatever is distinguishable, is separable by the thought or imagination. All perceptions are distinct. They are, therefore, distinguishable, and separable, and may be conceived as separately existent, and may exist separately, without any contradiction or absurdity."[28]

Bertrand Russell draws the full implication of the Humean doctrine for the theory of temporal succession:

There is no logical impossibility in the hypothesis that the world sprang into being five minutes ago, exactly as it then was, with a population that "remembered" a wholly unreal past. There is no logically necessary connection between events at different times; therefore nothing that is happening now or will happen in the future can disprove the hypothesis that the world began five minutes ago. Hence the occurrences which are *called* knowledge of the past are logically independent of the past; they are wholly analysable into present contents, which might, theoretically, be just what they are even if no past had existed.[29]

If past and present are merely extrinsically related, then the past is at best a construction built up from wholly present materials. From a Humean perspective, present percept-objects—memorative or imaginative images—provide us with the sole bases for construction of the past. The Humean tradition, moreover, with its inadequate phenomenology and its commitment to objectivist-reductionism, provides no basis for distinguishing imagining and remembering as differing intentional acts. Consequently, efforts of "mnemic" construction are ultimately arbitrary—so arbitrary that we are led to see that an external relation, in a significant sense, is no real, objective relation at all. The logic of external relations wreaks havoc with our temporal categories and renders our temporal concepts incoherent. In the end the logic of external relations, if consistently pursued, would issue first in a doctrine of sheer causal contingentism and ultimately in the skeptical solipsism of the present moment.[30]

The opposite tack, the attempt to interpret temporal succession within the context of a strong theory of internal relations that assimilates the causal order to that of logical implication, is the distinctive approach of rationalistic monisms and determinisms, whether materialistic (naturalistic) or idealistic (theological). The encompassing visions of Laplace's Demon and of Spinoza's God are virtually indistinguishable.

An often cited passage from Laplace—the passage alluded to in my earlier discussion of Bergson's criticism of Spencer's evolutionism—provides a classical statement of this rationalistic and monistic position: "An intellect which at a given instant knew all the forces acting in nature, and the position of all things of which the world consists— supposing the said intellect were vast enough to subject these data to analysis—would embrace in the same formula the motions of the greatest bodies in the universe and those of the slightest atoms; nothing would be uncertain for it, and the future like the past would be present to its eyes."[31] As noted earlier, for Laplace and his determinist confreres such as Spencer, the cause-effect relation is reduced to the relation of logical equivalence. If the logic of the cause-effect relation is that of mutual implication, it would follow that the difference between cause and effect, and with it, the difference between the past and the present, are merely conventions born of human limitations. The qualitative asymmetry of experienced durational succession would have to be banished to the inexplicable realm of "merely subjective" phenomena.

Bertrand Russell, now in a Laplacean humor, explicitly reduces the cause-effect relation to that of logical equivalence:

Causes, we have seen, do not *compel* their effects, any more than effects *compel* their causes. There is a mutual relation, so that either can be inferred from the other. . . . The apparent indeterminateness of the future . . . is merely a result of our ignorance. . . . Now, quite apart from any assumption as to causality, it is obvious that complete knowledge would embrace the future as well as the past. Our knowledge of the past is not wholly based upon causal inferences, but is partly derived from memory. It is a mere accident that we have no memory of the future. We might—as in the pretended visions of seers—see future events immediately, in the way in which we see past events. They certainly will be, and are in this sense just as determined as the past.[32]

Here, in the words of a most "secular" philosopher, we may detect the residue of a medieval, almost Boethian, theological vision. Complete knowledge would comprehend the essential identity and fixity of an eternal order, which order, from our finite perspective, inexplicably takes on an illusory visage of differentiation, novelty, and passage. *Sub specie aeternitatis,* past, present, and future coexist. Contingency gives way to necessity—to the necessity of the cosmic tautology.

Bergson's theory of durational succession and protomentalist metaphysics issue in a distinctive theory of causality—in a theory of causality that confronts us with a dialectical alternative to the theories propounded by either a psychologistic empiricism or by a logicistic and formalistic rationalism. Human reason does not confront the dilemma: *either* a Cratylean-Humean contingentism that, by entirely severing all continuity between past and present, eliminates objective causation, *or* a Parmenidean-Laplacean necessitarianism that, by cohesively welding past, present, and future, absorbs causation into timeless, static identity. There remains an Aristotelian-Bergsonian alternative: in the actual dynamic universe, genuine novelties emerge *both* as the syntheses of past antecedents *and* as the resolutions of an objectively indeterminate future.

Čapek provides a precise statement of this alternative:

In the temporal continuity of a real process of causation . . . the causal or "mnemic" influence of the past is not denied; but the present, though *codetermined* by the past, nevertheless contains an element of irreducible novelty. . . . In such a growing world every present event is undoubtedly caused, though not necessitated by its own past. For as long as it is not yet present, its specific character remains uncertain. . . . It is only its presentness which creates its specificity, i.e., brings an end to its uncertainty, by eliminating all other possible features incompatible with it. Thus every present event is by its own nature *an act of selection* ending the hesitation of reality between various possibilities. . . . As far as the future is concerned, it is the *future* and not a disguised and hidden present as in the necessitarian scheme. . . . But because it will not emerge *ex nihilo,* but from a particular present state, its general *direction* is outlined and thus possesses some general predictable features—the more predictable the larger the statistical complexes of the elementary events that are considered. Hence arises the possibility of practically accurate prediction of macroscopic events. (PICP 339–40)

For Bergson and Čapek the three durational *Exstases,* although dynamically interrelated, also must be qualitatively and tendentially differentiated. The *past,* as fully actualized and determinate, is objectively immortal. Aquinas, echoing Aristotle, correctly characterized this objective immortality of the past when he noted: "If the past thing is considered as past, that it should not have been is impossible, not only in itself *(non solum per se),* but absolutely *(sed absolute)* since it implies a contradiction."[33] The *present,* as immanently self-actualizing, is concretely active and subsistent in the strictest sense. The *future,* as presently unspecified but actualizable, is objectively indeterminate. Aristotle himself acknowledged the objective potentiality and indetermination of the future when he concluded that the law of the excluded middle does not apply to future situations *in their very futurity*.[34]

Bergson's theory of durational succession, his protomentalism, and his theory of causation thus converge in a generalized theory of agency. Each physical event, as a monadic process, *is* in virtue of its interior agency—an agency analogous to memory as an active synthesis. Each physical event, precisely because of the mnemic character of its interior agency, opens upon that causal past which provides the distinctive context for its own emergent novelty. The past, as causally efficacious, is immanent to the present; the present, although emergent and novel, conforms to, but is neither necessitated by nor identical with its past. The past is not so efficacious that it excludes the emergence of novelty; if it were, it would exclude its own character as past. The novelty of the present is not a novelty that excludes contextual conformation with the past, for the physical present is novel in virtue of, not in spite of, elementary memory. The physical present must be understood as novel in virtue of that immanent, mnemic agency that presently enlivens the past—a past that no longer exists in the concrete exercise of its own immanent agency. Further, each physical event immanently actualizes itself in the present, but also transcends itself in its causal, prospective relation to an emergent future. The Bergsonian physical event, like the Whiteheadian actual entity, is subject-superject, and Bergson is in fundamental agreement with Whitehead's comment: "An Actual entity is at once the product of the efficient past, and is also, in Spinoza's phrase, *causa sui*."[35]

2.3. Conclusion

I close with some suggestions concerning interrelations of evolutionary epistemology, durational metaphysics, and theoretical physics. These suggestions come to focus in terms of the following question: can such an epistemology and metaphysics, in their interdependence, provide a fruitful framework for present and future thinking in theoretical physics? I then identify a factor that continues to prevent the adoption of such a framework by contemporary physicists and philosophers of science: many, perhaps the majority, of theorists in these areas remain, despite recent developments in their disciplines, consciously or semiconsciously committed to the program of elementaristic reductionism and to a Spinozistic-Laplacean world view.

I first suggest that a distinction drawn from the Aristotelian-Thomistic tradition is relevant to an understanding of the protomental event in its character as subject-superject. This tradition distinguishes two forms of activity: *transitive activity,* which terminates in the perfecting of an entity other than the agent, and *immanent activity,* which terminates in and perfects an agent itself possessed of an active power. Sculpting, to cite a classic example, is a transitive activity that terminates in the production of the statue, that is, in the actualization of a new accidental form in the marble. Seeing, by way of contrast, is an immanent actualization of the subject's power of sight and terminates in the percipient's enhanced sensory knowledge of the object as visible, and not in the effecting of any new characteristic inherent to the material thing that is seen.

Process protomentalism may be regarded as a strong statement of the thesis that every finite actual occasion whatsoever possesses an emergent identity founded upon its proper *immanent activity.* This immanent self-actualization presupposes, as its necessary but not sufficient condition, a context of *transitive activity,* namely, the context provided by the causal efficacy of the past. Each finite actual occasion in its turn is *transitively* related to subsequent events, which come to be, *ab initio,* in their own dynamic right, as *immanently active.* Thus, in succession the transitivity of the past and the immanence of the present are inseparably linked in virtue of a mnemic prehensive function: even with respect to physical events, the "mere continuing of the before into the immediate after" evidences, and is uninterpretable apart from reference to, *both* the transitive efficacy of the past, *and* the immanent self-actualizing agency of the present.

In the older tradition, references to immanent activity occur solely within the context of the discussion of living beings and their vital powers. In discussing physical changes as physical, the tradition makes no reference to immanent activity. In relation to this premodern tradition, process protomentalism may be viewed *negatively* as the denial of the thesis that there exists a realm of merely physical entities—of entities devoid of immanent activity. *Positively,* process protomentalism generalizes: *to perdure, no matter how minimally, is to be immanently active.*

In the older tradition, cognitive relations—seeing, hearing, understanding, judging, etc.—are dynamic, internal, and asymmetrical correlations linking subject with object. In seeing, for example, the perceiving subject is transitively influenced by the external material thing and, as immanently active, intentionally becomes—cognitively internalizes—that object as visible. The intentional object, the thing as visible, is an abstractive simplification, but not a falsification, of the formal efficacy of the external material thing.

The same relational pattern—dynamic, internal, and asymmetrical—displayed in this premodern theory of intentionality is also evident, as Nicholas F. Gier indicates, in postmodern theories of intentionality. This relational pattern is intrinsic both to existential-phenomenological theories of intentionality and to Whitehead's theory of prehensive concrescence. This patterning of the subject-object correlation, moreover, provides, as it did within the older tradition, the basis, in the epistemologies of Whitehead and Merleau-Ponty, for moderate realism.

The notion of form is central to all three theories. As Gier notes: "These forms are not the hypostatized *eide* of Plato. Forms have no ontological status apart from their embodiment in experience. But, unlike conceptualist theory, these forms are inherent in experience itself and not merely conceptual. Phenomenologists maintain that experience is *already* pregnant with form, and this form is the *meaning of things,* not the meaning of a conceptual order." Gier further indicates that for Whitehead also, "meaning and form are 'objective,' but never in the sense of being apart from an intentional field or the prehensive unification of an actual occasion."[36]

In Čapek's analysis of Bergsonian protomentalism, the emergent agent-event as subject is dynamically, internally, and asymmetrically correlated with that causal past which constitutes the object-pole for the subject's mnemic synthesis. This is also, of course, the pattern of Whitehead's prehensive concrescence—a concrescence that displays the concrete universality of the Whiteheadian category "Creativity." Lewis S. Ford, commenting on Gier's study, notes that prehension, rather "than being simply identical with intentionality, . . . generalizes *both* intentionality *and* causality."[37] Ford's comment also applies to that mnemic synthesis which constitutes the focal center of Bergson's process protomentalism.

Bergsonian protomentalism, because of the prehensive character of even the most minimal mnemic synthesis, provides a generalized theory of immanent agency, intentionality, and causal efficacy. The theory, as a consequence, interprets evolutionary advance into novelty, at all levels, as displaying a variation-and-selective-retention pattern of development—as displaying a pattern analogous to Donald T. Campbell's generalization of the Darwinian natural-selection paradigm. Process protomentalism, especially when set within the context of Čapek's articulation of Bergson's biological theory of knowledge, generalizes Campbell's thesis: "Evolution—even in its biological aspects—is a knowledge process."[38] Within the context of process protomentalism, Campbell's thesis is expanded to read: evolution, not only in its biological aspects, but even at the level of physical events, is an intentional process interpretable in terms of a dynamic variation-and-selective-retention model.

Bergson's commitment to radical empiricism issues in a metaphysical generalization of the highest order. Ultimate *fact* and ultimate *generalization* are inseparable. As a radical empiricism, Bergson's systematic, as Čapek notes, is continuous with the basic thrust of the phenomenological movement, and Bergson's intuition of duration is a phenomenological *Wesensschau* of the profoundest significance:

> In his intuition Bergson certainly claimed to attain something more than his own private and momentary datum, something which, as he believed, can in principle be intuited by all persons who read and honestly try to understand his own introspective analysis. But what else is this "something" than the universal essence of duration, common, like the Heraclitean Logos, to all human percipients? . . . Duration, though always different and unique in its concrete content, must be *structurally* the same in different moments of one person as well as in different minds. In this sense the intuition of it is an insight into its *universal essence.* (BMP 68–69)

The Bergsonian theory of succession and durational agency, although experientially grounded in a most radical sense, is also universal in its significance and applicability. It is not to be confused with any abstract, univocal, formal concept of "becoming in general"; rather, it is to be characterized as a dynamic and concrete universal, analogically relevant to the interpretation of any instance of succession, whatever the concrete, qualitative content immanent to the individual process-event (cf. BMP 170–71). Thus, Bergson's metaphysics and Whitehead's philosophy of organism converge to a far greater extent than is generally recognized: for Bergson, as for Whitehead, "Creativity" is the universal of universals.

The Bergsonian systematic also provides a context for assimilating the distinctive suggestions of two leading contemporary theoreticians in the natural sciences. David Bohm indicates that the significance of internal relations—the mode of relatedness that links the events of the explicate order, through the rhythms of unfoldment and enfoldment, with the implicate whole—can ground, in a manner consonant with twentieth-century advances in physics, a culturally desirable, nonmechanistic vision of wholeness.[39] Ilya Prigogine indicates that the significance of dynamic and asymmetrical relations must be acknowledged by the contemporary physicist, that these relations bear an antireductionist and indeterminist import, and that focus on these relations and their implications provides a basis for the reintegration of scientific and humanistic concerns.

Bergson's process protomentalism asserts that the immanent agency of events concretely supports the dynamic, asymmetrical, and internal character of succession. Every event, as concretely real, is immanently active, and no event—not even one which is identified as an elementary physical entity—is merely an object, all "surface," devoid of subjectivity and internal significance. To think otherwise is to endorse the fallacy of simple location—to subscribe to the view that the individuality of an elementary physical entity "is based precisely on its ontological separation from other [merely externally related] simply-located entities" (BMP 309). The link between immanent agency and the dynamic ultimacy of succession grounds a contrary thesis: "The individuality and uniqueness of each event is based on its connection with its cosmical context" (BMP 309).

Can a revised physics—as not only Bergson and Whitehead, but also Bohm and Prigogine suggest—be grounded in such a vision? There can be no doubt that physics, throughout the classical period of its development, was the product of a "spectator-observer" viewpoint, which was and remains antithetical to the "agent-participant" perspective of Bergson and Whitehead. From the "spectator-observer" viewpoint, elementary physical entities are held to be qualitatively indistinguishable and externally related, not only to one another in the present of spatial juxtaposition, but also to their own individual past histories. Despite differences of individual history, it is maintained, each elementary entity remains qualitatively indistinguishable from the others and identical with the others in such quantitative features as charge and mass. The persisting influence of this viewpoint is evident in the vehemence with which even Karl R. Popper—a theorist who identifies himself as committed to an evolutionary epistemology, to realism, and to objective indeterminism—asserts that elementary physical entities "are physically completely identical *whatever their past histories*" and that such entities therefore cannot have anything like an inside state. Indeed, for Popper this fact constitutes the basis for rejecting any panpsychistic or protomentalist vision as being "so metaphysical (in a bad sense)," and as having "so little content," that it is virtually impossible to discuss its "explanatory power or . . . prospects."[40]

In Popper's assertions we may detect the residual influence of a major driving force operative throughout the modern period of Western scientific and philosophical development. Prigogine describes this force as belief in "the 'simplicity' of the microscopic";[41] I characterize it as the result of the Western mind's fascination with the force of elementaristic reductionism. This fascination was such that thinkers throughout the modern period seldom noted or effectually criticized its underlying assumptions—assumptions that James Collins specifies as follows:

The two main assumptions of *elementarism* are: *(a)* that the elements into which knowledge can be analytically resolved, are more authentic and reliable starting points than the complex wholes themselves, and *(b)* that the elements remain basically unchanged, whether taken in isolation or in composition. The elementarist viewpoint underplays the uniqueness of complex, cognitive wholes, and overlooks the profound modifications undergone by analytic units, upon their reintegration in the total context.[42]

Thinkers, whether classified as "empiricists" or as "rationalists" by subsequent generations of scholars, to the extent that they were committed to elementarism and to uncritical acceptance of its assumptions, were engaged in a fundamentally rationalistic and formalist program. Consequently, they were prone to fallacies of misplaced concreteness.

A strong synthetic urge also animated the modern period. The theoretical impulse issued in ever more encompassing formal systems of deterministic lawfulness. The Spinozistic-Laplacean vision became the guiding model for the development of Western notions of scientific rigor and intellectual responsibility. Although unlimited prediction and retrodiction were never more than theoretical possibilities, they nonetheless became essential elements of the scientific picture of the physical world. The possibility of such unrestricted prediction and retrodiction constituted, as Prigogine notes, "the founding myth of classical science" (FBB 214). Indeed, analysis and synthesis went hand in glove, for the task of elementaristic reduction became progressively governed by the conviction that analysis must terminate in elements that are amenable to a formalistic and deterministic synthesis.

As Prigogine further notes: "The situation is greatly changed today. It is remarkable that this change results basically from our better understanding of the limitations of measurement processes because of the necessity to take into account the role of the observer. . . . The incorporation of the limitation of our way of acting on nature has been an essential element of progress" (FBB 214). Developments in physics have themselves encouraged the abandonment of the "spectator-observer" perspective (the vantage point enjoyed by Spinoza's God and Laplace's Demon) and the humble acknowledgment of our own finite status as "agent-participants" in a cosmic and evolutionary context.

The consequent "increased limitation of deterministic laws means that we go from a universe that is closed, in which all is given, to a new one that is open to fluctuations, to innovations" (FBB 215), to a world view in which the irreversibility of durational succession is ultimate and with respect to which theoretical reversibility is a product of abstract idealizations.

The continuing hold of the Spinozistic-Laplacean vision upon the scientific imagination is evident when we contemplate expectable responses to the argument Čapek advances in defense of objective indeterminism in nature. Čapek argues that strict Laplacean determinism is incoherent. His argument may be formulated as follows:

1. If strict Laplacean determinism is an adequate and applicable theory, then the causal connection relating temporally successive events is a necessary relation.

2. If strict Laplacean determinism is an adequate and applicable theory, then all the features of a later event are logically deducible from any given temporally antecedent event and vice versa.

3. The sole necessary relation that permits the deducibility described in the consequent of premise 2 is the relation of logical equivalence.

4. If strict Laplacean determinism is an adequate and applicable theory, then the temporal relation of cause and effect is the timeless relation of logical equivalence.

5. It is logically impossible for a temporal relation to be a timeless relation.

6. *Therefore,* it is not the case that strict Laplacean determinism is an adequate and applicable theory. (Cf. BMP 106–11)

If a Humean objects to premise 3 and insists that the necessity linking cause and effect is logically contingent, then it is necessary that the characteristics of this nonlogical necessity be specified—and specified in a way that is immune to the charge of psychologism. Strict determinism requires unexceptionable causal laws. In what sense may a logically contingent relationship provide the basis for the universality and unexceptionable necessity of causal laws in both their predictive and retrodictive function? What basis still remains for distinguishing the cause-effect relation from that of equivalence? As David Sipfle comments in his perceptive review of Čapek's *Bergson and Modern Physics:* "It is ironic that Hume was a determinist since his analysis and redefinition of causality precluded any guarantee of unexceptionable causal laws, and it is precisely unexceptionable causal laws which are required by determinisms."[43]

Typically, the next step in the defense of Laplacean determinism is to charge that any theory that claims that the irreversibility of becoming is an inherent feature of temporal relations is subjectivistic and psychologistic. It is maintained that from the objective standpoint, that is, from the omniscient viewpoint that formal logic and mathematics alone afford, the qualitative distinctions of past, present, and future are irrelevant. Hence, although as premise 5 states, "It is *logically impossible* for a temporal relation to be a timeless relation," it is not logically impossible, but merely *psychologically difficult,* to acknowledge that a temporal relation can be a relation devoid of all overtones of irreversible becoming. All such overtones *logically* can be detached from the meanings we ascribe to basic temporal categories. When this "logical" feat is accomplished, qualitative becoming, with its irreversibility, stands revealed as mind- or consciousness-dependent. Einstein, Russell, and Adolf Grünbaum, along with many others, have endorsed this view.

Russell expresses this position as follows:

Both in thought and in feeling, even though time be real, to realize the unimportance of time is the gate of wisdom. . . . Every future will some day be past: if we see the past truly now, it must, when it was still future have been just what we now see it to be, and what is now future must be just what we shall see it to be. The felt difference of quality between past and future, therefore, is not an intrinsic difference, but only a difference in relation to us; to impartial contemplation it ceases to exist.[44]

Consequently, Russell can suggest: "we shall do better to allow the effect to be before the cause or simultaneous with it, because nothing of any scientific importance depends upon its being after the cause."[45]

With reference to the claims of Russell and others, Čapek comments:

All attempts to eliminate time [that is, qualitative and irreversible becoming] from physical reality create an unsolvable metaphysical enigma: How can timeless reality be transformed or unrolled into its illusory successive manifestations? Such attempts are even more grotesque when they are made by naturalistically minded scientists with behavioristic leanings; for consciousness, which is often dismissed by them either as an epiphenomenon or even as a mere remnant of "mentalistic metaphysics," is suddenly credited with the impossible achievement of transforming the static and becomingless character of reality into a successive and changing pattern. How can consciousness ever achieve such a magic trick, especially if it is associated with neural processes, that is, with a certain portion of the physical world which is allegedly timeless? (PICP 363–64)

The theory that irreversible, qualitative becoming is mind-dependent, and therefore irrelevant to physics, is clearly a theory that acknowledges that a persistent illusion is a brute fact, wholly uninterpretable in terms of the theory itself.

Bergson's speculative philosophy endorses the objective indetermination of the future, but it is not a version of acausalism. The formal patterns of the past *condition,* but *do not necessitate,* the present and future. Consequently, the general direction of the future is predictable, not at the level of individual elementary physical events, but at the level of statistical complexes of such events; thus "arises the possibility of practically accurate prediction of macroscopic events" (PICP 339–40). Unlike the mind-dependency theory, the Bergsonian theory does not resort to a category of persistent and uninterpretable illusion; it does square with recent developments both in the physical and biological sciences; and it is consonant with our fundamental experiential intuitions. It thus manifests a greater claim to the laurels of theoretical adequacy and experiential applicability.

Bergson's creative evolutionism—particularly as so skillfully explicated and expanded by Čapek—rejects the necessitarian schemes of mechanistic determinism and extrinsic finalism. It affirms the reality of objective potencies inherent to nature and the objective indetermination of an open future. In the evolving community of protomental events, each agent-event, in its unique exercise of the immanent teleology of organism, both embodies and transcends its effectual past. Further, each actual entity, in bringing the possibilities of the present to fruition, also effects the present resolution of an ever indeterminate future. The becoming of continuity and the emergence of novelty are the twin facets of a present synthesis which, in its very presentness, is both mnemic and creative. We thus may safely dismiss Einstein's fears concerning God's playing dice with the universe, and we may yet discover a profound inspiration in another of Einstein's claims: the Lord of History—although He is not to be identified with Spinoza's God—is nonetheless not malicious, but He is indeed tremendously and magnificently subtle!

Abbreviations of Frequently Cited Works

BMP   Milič Čapek, *Bergson and Modern Physics* (Dordrecht, Holland: D. Reidel, 1971)

CE   Henri Bergson, *Creative Evolution,* trans. A. Mitchell (New York: Modern Library, 1944)

CM   Henri Bergson, *Creative Mind,* trans. M. L. Andison (Totowa, NJ: Littlefield and Adams, 1965)
FBB   Ilya Prigogine, *From Being to Becoming* (San Francisco: W. H. Freeman, 1980)
FP   Herbert Spencer, *First Principles* (New York and London: D. Appleton, 1910)
OE   Henri Bergson, *Oeuvres,* ed. André Robinet (Paris: Presses Universitaires de France, 1959)
PICP Milič Čapek, *The Philosophical Impact of Contemporary Physics* (New York: Van Nostrand-Reinhold, 1961)

Notes

1. John Dewey, *The Influence of Darwin on Philosophy* (Bloomington: Indiana University Press, 1965), 1–2.
2. Cf. Milič Čapek, "Ernst Mach's Biological Theory of Knowledge," *Synthese* 18 (1968):149; and Čapek's "The Significance of Piaget's Researches on the Psychogenesis of Atomism," in *Boston Studies,* vol. 8, ed. R. C. Buck and R. S. Cohen (Dordrecht, Holland: D. Reidel, 1970), 448.
3. Čapek, "Ernst Mach's Biological Theory of Knowledge," 148.
4. Ludwig Wittgenstein, *Tractatus logico-philosophicus,* trans. D. F. Pears and B. F. McGuinness (London: Routledge & Kegan Paul, 1961), 49; subsequent references to the *Tractatus* cite proposition numbers and are included in the body of the text.
5. A. J. Ayer, *Language, Truth, and Logic* (1946: reprint, New York: Dover, n.d.), 84; the emphases are mine.
6. Milič Čapek, *Bergson and Modern Physics* (Dordrecht, Holland: D. Reidel, 1971), 54; hereafter referred to as BMP.
7. In Čapek's view, philosophers of a formalistic or rationalistic orientation—because of their conscious or semiconscious metaphysical commitments, because of their unexamined assumptions, or because of their limited interests—have been, and continue to be, hostile or indifferent to evolutionary and genetic epistemologies (cf. BMP 7, 63). Stephen Toulmin provides a similar assessment in his *Human Understanding* (Princeton: Princeton University Press, 1972), 1.58 and 116.
8. Toulmin, *Human Understanding,* 1.13; cf. 1.2, 4, 25.
9. For Popper's recent contributions to evolutionary epistemology, see his *Objective Knowledge: An Evolutionary Approach* (Oxford: Clarendon, 1972); his *Unended Quest: An Intellectual Autobiography* (London: Fontana/Collins, 1976); and a work he coauthored with John C. Eccles, *The Self and Its Brain* (New York: Springer-Verlag, 1977). Campbell's contributions include: "Variation and Selective Retention in Socio-Cultural Evolution," in *Social Change in Developing Areas: A Reinterpretation of Evolutionary Theory,* ed. H. B. Baringer et

al. (Cambridge, Mass.: Harvard University Press, 1963), 19–49; "Blind Variation and Selective Retention in Creative Thought and in Other Knowledge Processes," *Psychological Review* 67 (1960):390–400; and "Evolutionary Epistemology," in *The Philosophy of Karl Popper,* ed. P. A. Schilpp (LaSalle, II: Open Court, 1974), 1.413–63.

10. Campbell, "Evolutionary Epistemology," 1.413.

11. Milič Čapek, *The Philosophical Impact of Contemporary Physics* (New York: Van Nostrand-Reinhold, 1961); hereafter referred to as PICP.

12. Herbert Spencer, *First Principles* (New York and London: D. Appleton, 1910), 70–72; hereafter referred to as FP.

13. For Čapek's analysis of the manner in which Spencer's thought, although representative of the transition from eighteenth-century dynamism to modern energism, displays the main features of the corpuscular-kinetic view, see PICP 100–02.

14. For some indication of the character of the convergence of Spencer's empiricism and Kant's rationalism, see FP 148.

15. Spencer defends the practical equivalence of subjective correlation and ultimate, objective reality in FP 141–45; cf. Čapek's comments in PICP 102.

16. This commitment to realism grounds Spencer's conviction that the truths expressed in fundamental logicomathematical and scientific sentences are objective: the truth of such sentences remains independent of the person who asserts them and of the spatiotemporal context in which they are asserted.

17. Cf. Henri Bergson, *Oeuvres,* ed. André Robinet (Paris: Presses Universitaires de France, 1959), xiv–xx, 1541–43; hereafter referred to as OE. When citing English translations of Bergson's works, I shall also indicate, with an equal sign, the corresponding page(s) in OE. Cf. also Bergson's *Creative Mind,* trans. M. L. Andison (Totowa, NJ: Littlefield and Adams, 1965), 11–17, 26–29, 91–95 = OE 1253–59, 1268–70, 1331–35 (hereafter referred to as CM); and his *Écrits et paroles,* ed. R. M. Mossé-Bastide (Paris: Presses Universitaires de France, 1959), 1.204, 2.238–40, 294–95, 3.456.

18. Bergson, *Écrits et paroles,* 2.295; my translation.

19. Cf. Bergson, *Creative Evolution,* trans. A. Mitchell (New York: Modern Library, 1944), 45 = OE 526; hereafter referred as to CE.

20. With reference to this text, it is interesting to note the pointed criticisms that C. S. Peirce provides with respect to Spencer's mechanistic evolutionism. See *Collected Papers of Charles Sanders Peirce,* ed. Charles Hartshorne and Paul Weiss (Cambridge: Harvard University Press, 1965), 6.15–16. The parallels with Bergson's critique of Spencer's evolutionism are most striking.

21. Reichenbach provides a strong statement of the biological theory of knowledge: "The conception of a corporeal substance, similar to the palpable substance shown by the bodies of our daily environment, has been recognized as an extrapolation from our sensual experience. What appeared to the philosophy of rationalism as a requirement of reason—Kant called the concept of substance a synthetic a priori—has been revealed as being the product of a conditioning through environment. The experience offered by the atomic phe-

nomena makes it necessary to abandon the idea of a corporeal substance and requires a revision of the forms of the description by means of which we portray physical reality. With the corporeal substance goes the two-valued character of our language, and even the fundamentals of logic are shown to be the product of an adaptation to the simple environment into which human beings were born." (*Atom and Cosmos,* trans. E. S. Allen [New York: G. Braziller, 1957], 288–89; cf. BMP 65–67.)

22. Bergson, *Matter and Memory,* trans. N. M. Paul and W. S. Palmer (London: George Allen and Unwin, 1962), 17 = OE 179.

23. This section incorporates materials from, and elaborates upon issues posed in, my articles, "Čapek, Bergson, and Process Proto-Mentalism," *Process Studies* 11, 3 (Fall 1981):180–89, and "Popper's Critique of Panpsychism and Process Proto-Mentalism," *Modern Schoolman* 59 (May 1982):233–54, especially 234–43.

24. Čapek explicitly rejects the frequent misrepresentation of Bergson as a dualist in the Cartesian tradition (cf. BMP 30). His central treatment of process protomentalism is in BMP 302–11, part 3, chap. 14, "Physical Events as Proto-Mental Entities—Bergson, Whitehead and Bohm." For my defense of the anti-Cartesian, pluralistic character of Bergson's metaphysics, cf. "Bergson's Dualism in *Time and Free Will,*" *Process Studies* 4, 2 (Summer 1974):93–106.

25. Bergson, *Duration and Simultaneity,* trans. Leon Jacobson (Indianapolis: Bobbs-Merrill, 1966), 48–49.

26. Gottfried Wilhelm Leibniz, *Theoria motus abstracti seu rationes motuum universales a sensu et phaenomenis independentes,* in *Die philosophischen Schriften von Gottfried Wilhelm Leibniz,* ed. C. I. Gerhardt (Berlin: Wiedmann, 1875–1890; unchanged reprint, Hildesheim: Georg Olms, 1965), 4.230. For Čapek's extended analysis of the relation of Leibniz's thought to that of Bergson and Whitehead, cf. his "Leibniz on Matter and Memory," in *The Philosophy of Leibniz and the Modern World,* ed. Ivor Leclerc (Nashville: Vanderbilt University Press, 1973), 78–113, especially 96–104 and 111–13.

27. Alfred North Whitehead, *Process and Reality,* Corrected edition, ed. D. R. Griffin and D. W. Sherburne (New York: Free Press, 1978), 3.

28. David Hume, *A Treatise of Human Nature,* ed. L. A. Selby-Bigge (Oxford: Clarendon Press, 1868), 634.

29. Bertrand Russell, *The Analysis of Mind* (London: George Allen and Unwin, 1921), 159–60.

30. Čapek notes that Russell, in *Human Knowledge, Its Scope and Limits* (New York: Simon and Schuster, 1948), 90, "does not stop short of the most extreme form of 'the fallacy of simple location in time', . . . when he claims that not only is a man private from other people, but he is also private from his own past" (BMP 344). With reference to the logical possibility "that the world sprang into being five minutes ago, exactly as it then was, with a population that 'remembered' a wholly unreal past"—a possibility Russell also asserts in *Human Knowledge,* 212—Čapek comments: "Russell . . . concedes that nobody takes such supposition seriously. But . . . if nobody takes seriously

what cannot 'logically' be refuted, then there must be something radically wrong with a logic of this kind" (BMP 344).

31. Pierre Simon de Laplace, *Introduction à la théorie analytique des probabilites,* in *Oeuvres completes* (Paris, 1886), vi; cited and translated by Čapek, PICP 122.

32. Bertrand Russell, *Our Knowledge of the External World* (London: George Allen and Unwin, 1921), 234–35.

33. *Summa Theologica,* 1.25, *ad* 1.

34. *De Interpretatione,* 9.18a34–19b4.

35. Whitehead, *Process and Reality,* 150.

36. Nicholas F. Gier, "Intentionality and Prehension," *Process Studies* 6, 3 (Fall 1976):305.

37. Ford's comment is quoted by Gier, 205–06.

38. Cf. the discussion at the beginning of this chapter and note 10, above.

39. This is particularly true when Bohm's vision of wholeness is interpreted as *prehensive wholeness* along the lines articulated by David Ray Griffin in his paper in this volume. Griffin's suggestions are developed in reference to the latent Spinozism inherent to the conception of implicate wholeness as an *underlying* and *grounding,* rather than *prehensive,* wholeness. Although Bohm rejects mechanism, he nevertheless seems to remain open to the acceptance of a version of determinism grounded in a strong theory of internal relations— that is, in a theory that endorses the internality of ultimate relations, but that rejects the view that ultimate relations are also dynamic and asymmetrical. Bohm indicates that *explicate relata* are internal to one another solely in virtue of their common internal relations to the *implicate whole.* Bohm also indicates, at least in some passages, that the implicate whole, as grounding the explicate relata, is not interpretable in terms of relations that are internal in the weak sense, that is, in the sense that they are not only internal, but also dynamic *and* asymmetrical.

40. Popper, *The Self and Its Brain,* 71.

41. Ilya Prigogine, *From Being to Becoming* (San Francisco: W. H. Freeman, 1980), xiii; hereafter referred to as FBB.

42. James Collins, *A History of Modern European Philosophy* (Milwaukee: Bruce, 1954), 314.

43. David Sipfle, "Review of Milič Čapek's *Bergson and Modern Physics,*" *Process Studies* 2, 4 (Winter 1972):309–10.

44. Russell, *Our Knowledge of the External World,* 226.

45. Russell, *Mysticism and Logic* (New York: Norton, 1929), 21–22.

3. Pete A. Y. Gunter

DYNAMIC, ASYMMETRICAL, INTERNAL RELATIONS:
SOME QUESTIONS FOR ANDREW BJELLAND

(1) I would like to thank Andrew Bjelland for raising the question of biological theories of knowledge. In a conference devoted to physics, it is quite likely that this question would be overlooked. And yet, it is essential and unavoidable. One wonders how many of the physicists or philosophers who participate in this conference would accept, or defend, a biological epistemology.

For a biological theory of knowledge, concepts are tools shaped by the organism through interaction with its environment. Man's intellectual evolution would then be the development of a superior set of tools (better adapted, more powerful) at the service of human survival. This all seems plausible. But, like any comprehensive view, it has its difficulties. I would like to stress two of these, which were urged by Bertrand Russell and G. E. Moore just after the turn of the century.

The first is a charge of vicious circularity. One accepts evolutionary theory, and on this basis accepts a biological theory of knowledge. But for such a theory of knowledge, evolutionary theory is itself only a set of tools for adapting to experience. Has one, then, committed a *petitio principii,* supporting a biological epistemology *via* evolutionary theory and evolutionary theory *via* a biological epistemology? There are many ways of putting this objection. But are we sure that, however the objection is put, we can formulate our biological theory of knowledge so as to void them?

Another, and closely related, question concerns the status of purely formal concepts (i.e., logic and mathematics) in a biological theory of knowledge. On this theory—and the theory has appeared least plausible at this point, especially to pure mathematicians—mathematics and logic provide the most general form of the human adaptation to nature and are thus, whatever Pythagorus or Plato may have thought of them, survival instruments. But if this were so, the objection might run, then our most basic mathematics, even the principles of identity and contradiction, could at least conceivably be changed. We would not accept such a change, however, and could not. Moreover, logic and mathematics have a strictness and precision that biological conceptions do not. In short, can we defend the basic, strict, unchanging character of mathematics and logic, given a biological epistemology?

(2) I would also like to thank Bjelland for making clear just what all philosophers of process have in common and what all antitemporalist philosophers deny: that is, that time is a dynamic, asymmetrical, internal relation. It is significant that on the basis of this distinction not only philosophers of process but phenomenologists, pragmatists, Aristotelians, and Thomists stand in agreement on the same fundamental issue.

This agreement, however, masks a fundamental issue on which there may be significant disagreements. It is this. Granted that temporal relations are asymmetrical and internal, does this asymmetry spring from the past's internal relations with the present or from the present's internal relations with the past? In other words, does the past, or some aspect of it, literally move into the present, grounding the observed asymmetry? Or does the present, once it is extant, "reach back" and infiltrate the past? In the recent conference on Henri Bergson and modern science at Galveston, Charles Hartshorne repeatedly stressed the necessity of the former option and the unintelligibility—even the internal contradictoriness—of the latter.

In sum: something needs to be said about this issue.

(3) The "process consensus" that time is a dynamic, asymmetrical, internal relation may help us, in a general way, to resolve a serious philosophical dilemma whose "horns" are represented by Hume at one extreme and Spinoza at the second. That is, the process consensus should:

1. make it possible to refute or otherwise undermine the skepticism of sensationalist doctrines of time as an *external* relation (Hume, and Russell—in empirical moods).

2. make it possible to overcome the paradoxes of entailment theories of time as a *static internal relation* (Spinoza, and Russell—in rationalist moods).

Would Bjelland be willing to take on the refutation of Russell's claim that no one could tell if the world were created five minutes ago, all its people having false memories? It would not be enough, to do this, simply to say that Russell's notion of time is wrong.

Finally: on the view that time is a dynamic, asymmetrical, internal relationship, can we say, simply, that time is "continuous"? Or doesn't creativity entail a rupture with the past, and hence discontinuity?

One wonders to what extent these questions (biological epistemologies and their internal consistency, the problems of past-present-past relations, of the recent and the distant past, of continuity-discontinuity) will be of interest to scientists.

4. Andrew G. Bjelland

RESPONSE TO PETE GUNTER

(1) Pete Gunter's first set of questions deals with two objections of Russell and Moore. These objections have considerable force when they are directed toward a biological epistemology, such as Spencer's, that is grounded in a mechanistic interpretation of evolution and an associationist theory of mind and that stresses the link between truth and biological utility. The objections, however, fail to come to grips with biological epistemologies—such as those of Bergson and, more recently, of Karl Popper—that are grounded in a *metaphysics* of emergent evolutionism.

I believe that Popper is right when he claims that evolutionary theory is a nonfalsifiable *metaphysical research program* that provides a fruitful framework both for scientific inquiry and for epistemological reflection. In my view, evolutionary metaphysics and biological epistemology are intimately and dialectically interrelated. This interrelation, however, because of its dialectical character does not involve a vicious circle.

From the perspective of Bergson's biological epistemology, the formal categories of the Newtonian-Euclidean world view are, as Gunter puts it, "a superior set of tools . . . at the service of human survival." These objectifying categories, however, do not set the limits of the intelligible. Consequently, truth need not be equated with survival value. Survival value provides a test for the range of the applicability of our formal concepts, but does not constitute the definition of truth.

For Bergson and other process thinkers, the person's reflective participation in, and embodiment in terms of, the dynamic, asymmetrical, and internal relations characteristic of process constitute the ultimate grounds of intelligibility. The theoretical impulse arises in the concrete context of durational embodiment, but it is not reducible to biological categories.

Personal agency and process, although intelligible, elude objectification. Agency and process are neither derivable from, nor interpretable in terms of, univocal, formal concepts. *Intuition* of the dynamic patterns of agency and process—intuition that begins as a vague awareness of the inadequacy of our formal concepts but that comes to fruition only through dialectical reflection—transcends the frames of our pragmatically instrumental, stabilizing categories. These dynamic patterns are experiential "givens." The foundational significance of these "givens" is captured, however, only through careful phenomenological analyses, coupled with sustained critical reflection on the act-object contrasts and ambiguities that have haunted Western philosophy since the dawn of the modern era.

With respect to Gunter's queries concerning the status of the principles of identity and contradiction, it seems to me that process thinkers should emphasize that these are not merely formal, logical principles, but are at root metaphysical principles.

Process philosophy, as Whitehead noted, is to a considerable extent a reclamation of premodern modes of thought. (These modes of thought, to be sure, are not only reclaimed—their significance is transformed when they are set within the context of postmodern cosmology.) Modern philosophy, in both its rationalist and empiricist wings, is marked by the tendency to employ logical categories to set the limits of all possible experience. In the premodern Aristotelian view, the ontological grounds the logical.

In the older tradition, basic principles of thought are viewed as having their grounding in the Logos of the real, in the interrelationships of agency which constitute the community of being. The cognitive subject's reflective participation within the community of being served as the experiential warrant for the significance and universality of fundamental principles—for their character as *per se notum,* dialectically defensible principles.

In the postmodern period, we are well acquainted with the extent to which formal systems admit of expansion and with the manner in which new modes of formalism emerge and evolve. From the Bergsonian perspective, the Logos of the real is manifested in the dynamic, asymmetrical, and internal relations of durational process, and our principles of interpretation, both dynamic and formal, are ultimately validated by our reflective participation in that world-process. A proper understanding of the emergence and evolution of formal systems demands appeal to metalogical and metamathematical principles—to principles which are invariant, but relational, precisely because they are grounded in the dynamic patterns of this process.

(2) Gunter next poses the question: Is the asymmetry of succession grounded in the internality of the past to the present, or of the present to the past? He then cites Hartshorne's objection that the present cannot be internal to the past. Hartshorne's objection is voiced in defense of the claim that the present must be truly novel and therefore not precontained in, and deducible from, its antecedents.

The asymmetry of succession is grounded in the prehensive agency of the present event. As novel, mnemic synthesis, the present event both internalizes and transcends the determinate past. The present event is truly emergent and is not predetermined by the past. Bergson and Hartshorne are in total agreement on this point.

Durational succession is intelligible solely in terms of a *sui generis* qualitative differentiation of past from present which depends on the mnemic, yet novel, survival of the past in the present. The past does not precontain the present event, nor does the present event merely reiterate the past. Mere precontainment and mere reiteration are limit concepts; the realization of either would entail the collapse of succession into identity. These issues are dealt with in further detail in section 2.2 of my paper.

(3) Gunter requests a refutation of Russell's claim that nothing can disprove the hypothesis that the world began five minutes ago, with everyone having false memories. The following comments may not satisfy Gunter's request. They may remain too close to saying that Russell's notion of time is simply wrong. I hope, however, that my comments at least provide a sketch for a refutation.

The "logic" in terms of which nothing can disprove Russell's hypothesis is, as I indicate in section 2.2, the atomistic logic of external relations. Within the context of logical atomism, Russell's claim is as unassailable as it is trivial. In this context, all efforts of mnemic construction are equally arbitrary—the logic of external relations is inextricably linked with the solipsism of the present moment. Thus, the "logical" underpinnings for Russell's claim ground an untenable epistemological position, providing sufficient reason for refusing to take the hypothesis seriously and for concluding that the logic of external relations is an inadequate instrument for the analysis of temporal notions and terms. (Further criticism would involve detailed discussions of the objective immortality of the past and of puzzles associated with "false memories.")

Gunter's final questions concern the continuity of succession and the extent to which creativity entails a rupture with the past, and hence discontinuity. I believe that my response (2), above, and the discussion in section 2.2 of my paper constitute an adequate response to these questions.

5. Patrick Hurley

TIME IN THE EARLIER AND LATER WHITEHEAD

The concept of time plays a central role in Whitehead's writings. Not only did Whitehead write two separate essays entitled "Time" and two others on space-time, but all his major works, beginning with the 1905 memoir *On Mathematical Concepts of the Material World,* make frequent reference to the subject. Time in Whitehead, though, is a difficult subject, and it would be a mistake to surmise that a coherent account of his position can be obtained by simply fusing together the various references to time as they occur in these works. Part of the trouble is that Whitehead uses the word *time* in several different senses. In some passages he speaks of time in the concrete and in others of time in the abstract. And then there is time as it occurs in science and time as we experience it directly. So any attempt to understand Whitehead's philosophy of time requires, at the very least, that we distinguish the various senses in which the word is used and determine how they relate to one another.

A second problem, one that is readily apparent to anyone who has made even a cursory survey of Whitehead's work, is that the writings reflect a distinct evolution of philosophical insight. The early works deal primarily with logic, mathematics, and the philosophy of science, whereas the later ones extend into metaphysics and the philosophy of civilization. With this growth in perspective, there is a corresponding deepening of appreciation for the importance of time. Thus, to understand Whitehead's philosophy of time, we must also come to grips with how time in the early writings relates to time in the later ones.

Any successful effort to deal with these difficulties requires that we locate Whitehead's treatment of time within the context of his overall philosophical objective. It is my contention that there is such an objective and that it remains the same from the very earliest of his works (including the purely mathematical treatises) through the very latest. This objective can be seen most clearly through an examination of Whitehead's methodology. Accordingly, in the account that follows, I will first take up a brief explanation of what that methodology is and then turn to how it bears upon the question of time, considering first the early writings and second the later ones.

5.1. Whitehead's Methodology

To obtain an appreciation for what Whitehead tried to accomplish in the course of some fifty years of philosophical productivity, we must inquire into the intellectual milieu that prevailed at Cambridge University during the latter part of the nineteenth century. Whitehead entered Cambridge in 1881 and studied pure and applied mathematics exclusively. Three years later he selected Maxwell's classic *Treatise on Electricity and Magnetism* as the subject of his fellowship dissertation. Thus, more precisely, we must inquire into the scientific intellectual milieu at Cambridge during those years. The principal figures in the Cambridge scientific community during the latter part of the nineteenth century were Clerk Maxwell, Kelvin, Stokes, and J. J. Thomson, all of whom shared a fundamental belief about scientific methodology. This was the conviction that important scientific discoveries can be made only through the construction of models. Kelvin went so far as to say that the meaning of "Do we or do we not understand a particular point in physics?" is "Can we make a mechanical model of it?"[1] In other words, for Kelvin, scientific explanation was equivalent to constructing a model that illustrated the phenomenon to be explained.

Kelvin made models for all sorts of phenomena: static electricity, atoms, the interaction of fundamental particles, electromagnetic induction, and the structure of crystals. Stokes built a highly elaborate mathematical model for the luminiferous ether. Maxwell built models to explain the rings of Saturn and the phenomena of electrostatic and magnetic force, as well as a highly imaginative vortex model to explain electric currents, electromagnetic induction, and the rotation of the plane of polarized light in magnetic fields. The electromagnetic field, which is the crowning achievement of the *Treatise,* is a highly abstract model that interprets the interaction of all electromagnetic phenomena. Finally, J. J. Thomson, discoverer of the electron, attempted to identify the structure of the atom through the construction of models, but none proved entirely successful.

Some of the models produced by these men were merely mechanical contraptions that illustrated the internal operation of the phenomenon to be explained. But others were highly imaginative, mental constructions that could never really exist in the physical world. Maxwell's were primarily of the latter sort. Joseph Larmor recalls that "with Maxwell the scientific imagination was everything. . . . Maxwell reveled in the construction and dissection of mental and material models." These models laid the foundation for "the bold tentative flights into the unknown which in Maxwell's work turned out to be so successful."[2]

The reason Maxwell, Kelvin, Stokes, and Thomson were so convinced of the power of model construction is that a model, in virtue of its internal coherence, exposes the interconnectedness of phenomena that, prior to the application of the model, were known only separately. For example, prior to Maxwell's invention of the electromagnetic field, the phenomena of light, electricity, and magnetism were known only separately and were studied as separate branches of physics. The electromagnetic field succeeded in bringing them together in terms of a single explanatory scheme. The model accomplishes its function through certain concepts that are linked together by a network of mathematical expressions. When the phenomena of light, electricity, and magnetism are interpreted as *instances* of these concepts, they too become linked together.

Through his work on Maxwell's *Treatise,* Whitehead became intimately familiar with the methodology of model construction. In *Science and the Modern World* (henceforth SMW), he referred to it as the method of invention:

The greatest invention of the nineteenth century was the invention of the method of invention. A new method entered into life. . . . We must concentrate on the method in itself; that is the real novelty that has broken up the foundations of the old civilization. (SMW 96)

Whitehead was impressed with the method of model construction for the same reason that the Cambridge scientists were: its ability to interconnect disparate classes of phenomena. In fact, I believe he was so impressed with it that he used it himself in all his major works. For example, in *Universal Algebra,* Whitehead developed a generalized concept of space to unite the various algebras. This generalized concept of space is a model. We can call it a "cognitive model." When elements of the various algebras are seen as instances of concepts in the model, they become linked together just as light, electricity, and magnetism are linked together through Maxwell's electromagnetic field. In *Principia Mathematica* (co-authored with Bertrand Russell), the same objective was extended to the whole of mathematics, but the model in this case was constructed from materials supplied by logic. In *Mathematical Concepts of the Material World,* Whitehead constructed six mathematical models that depict possible ways of interconnecting the fundamental concepts of physics. In *An Equiry concerning the Principles of Natural Knowledge,* he adapted the method of class constructionism used in *Principia Mathematica* to produce a cognitive model that would interconnect our perception of nature with our scientific knowledge of nature. By so doing, he hoped to avoid what he called the "bifurcation" of nature—the separation of our scientific knowledge of nature from our everyday expression of this knowledge. In *The Concept of Nature,* the project begun in the *Principles of Natural Knowledge* was rendered in a more popular vein. Finally, in *Process and Reality,* Whitehead wanted to construct a "scheme of ideas" that would succeed in interconnecting all that is given for "thought, for perception, and for feeling." This scheme of ideas is simply another name for a cognitive model.

If we grant that Whitehead did indeed use Maxwell's method of model construction in all his major writings, it follows that there is an interpretive requirement that we must consistently adhere to when approaching these writings. We must distinguish language that is used to describe the model and its construction from language that describes the data that the model is intended to interconnect. The former may otherwise be termed "systematic" language, the latter, "presystematic" language.³ If the two are confused, there can be no hope of making any sense of Whitehead's position. In reference to the problem of time, systematic language about time will always refer to something relatively abstract. This follows simply from the fact that the model is composed of abstract concepts. On the other hand, language used to describe the data given for interconnection may be either abstract or concrete, depending on Whitehead's particular purpose at the time he applied the model.

One final comment needs to be made about what Whitehead takes these data to be. It appears that, in all his writings, these data are selected from the content of subjective awareness. In other words, there is a distinctly subjectivist orientation to Whitehead's employment of the method of model construction. This subjectivist orientation I see as having arisen from the influence of Kant. Whitehead tells us himself that while he was a student at Cambridge, he knew by heart parts of Kant's *Critique of Pure Reason* (AB Notes 13; see list of abbreviations at end of chapter), and Russell says that Whitehead "always had a leaning toward Kant."⁴ This leaning never went so far as to cause Whitehead to think that the world outside the mind is unknowable, but he was always committed to the idea that we should begin doing philosophy by analyzing the content of subjectivity. In the early writings, Whitehead makes no attempt to establish a link between the content of subjectivity and the world outside the mind, and as a result his philosophy exhibits a certain tone of aloofness. In the later works, however, he uses model construction precisely for this purpose. The theory of prehension shows how the content of subjective awareness arises from the settled world of experience outside the mind.

5.2. Time in Whitehead's Early Writings

The first major work that contains a treatment of time is *Mathematical Concepts of the Material World* (henceforth MC). In that work, Whitehead develops six separate models to interconnect the fundamental entities of the physical world. These fundamental entities are termed "ultimate existents" and, depending on the particular model, consist variously of instants of time, particles of matter, points of space, particles of ether, and lines of force. In each case, the model itself consists of three kinds of components: the fundamental relations, the definitions, and the axioms. The fundamental relations, in turn, include the essential relation, the extraneous relations, and the time relation. The essential relation establishes the basic character of the model, the extraneous relations supplement the essential relation in various ways (for example, in Concepts I and II, it is used to account for motion), and the time relation is a dyadic relation that is used purely for the purpose of ordering the instants of time.

MC is interesting in regard to Whitehead's philosophy of time for a number of reasons. First, in that work Whitehead always speaks of time as composed of *instants*. There is no trace of the idea expressed some years later that real time must always involve a durational component and that instants are only abstractions from durations. At this point Whitehead sees no problem with the notion that instants of time are part of the data immediately given for philosophical analysis, and he proceeds with the construction of a mathematical model that will link them with the other basic components of physics.

Another interesting feature of MC arises from Whitehead's treatment of what he terms the "objective reals." The objective reals are the ultimate existents after the instants of time have been subtracted out. Because Whitehead calls these entities "objective," does this mean that the points of space, particles of ether, and so on are taken as existing independently of the mind? Not at all. Whitehead says, "We have no concern with the philosophic problem of the relation of any, or all, of these concepts to existence" (MC 13). In other words, the objective reals *might* exist independently of the mind, but Whitehead is not committed to the idea. In what sense, then, are the objective reals "objective"? The answer is that they are objective in the same sense that nature is objective for Kant: they are presented as objects of sense experience. But if the objective reals are objective only in this sense, what can we say about the instants of itme, which are *not* taken as objective? Here the answer seems to be that, at this point, Whitehead might not even consider time to be real. Whitehead's Cambridge colleague McTaggart held to the unreality of time, and in his many conversations with Whitehead he might have persuaded him to adopt such a view, at least provisionally.

The fact that time is something quite different from space is supported, at this point in Whitehead's development, by the idea that a special time relation is needed to order the instants of time, but that no special space relation is needed for the points of space. Also, points of space are part of the objective reals, but not so the instants of time. Thus, at this point, there is no trace of the idea that space and time are components of a single manifold, as they were for Einstein in 1905. Furthermore, the fact that the time relation is always dyadic precludes the possibility of multiple time systems.

Since Whitehead was unfamiliar with Einstein's work at this point, does this mean that he was a Newtonian? Concepts I and II are called Newtonian because space and time are completely separate from one another. The field of the essential relation includes points of space, and instants of time are connected with the points of space via the extraneous relations. The remaining four concepts are termed "Leibnizian" because the field of the essential relation includes both points of space and instants of time and because both space and time as a whole are conceived as arising from the relationships between the other entities in the field. The alternative approaches of these cognitive models would therefore suggest that, at this time, Whitehead was committed neither to the Newtonian nor the Leibnizian viewpoints. Whichever viewpoint was preferred, Whitehead showed how to construct a model that would weld the fundamental entities into a cohesive whole.

During the five years following the presentation of *Mathematical Concepts of the Material World,* Whitehead was taken up almost exclusively with the writing of *Principia Mathematica.* Nothing in this massive treatise has any direct bearing on either space or time, but it does have an indirect bearing, since the technique of class constructionism developed and employed in that work provided the basis for the method of extensive abstraction, which was presented for the first time in *The Relational Theory of Space* (RTS). This treatise was read to the First Congress of Mathematical Philosophy, held in Paris in 1914. RTS is, as the title implies, a work about space and not time. Nevertheless, Russell tells us that Whitehead had, as of its writing, extended the method of extensive abstraction to time even though he did not include this treatment in the published version.[5]

The objective of RTS is to construct a model, using the class logic of *Principia,* to establish a bridge between what Whitehead called "apparent objects" (e.g., green trees, sounds, colors) and the points, lines, and planes of perceptual geometry. Alternately, the model is also used to provide a bridge between what he calls "physical objects" (e.g., atoms, molecules, and electrons) and the points, lines, and planes of physical geometry. This dual application of a single logical model is accomplished by having the concepts of the model serve as variables. When, in one mode of application, apparent objects and the points, lines, and planes of perceptual geometry are interpreted as instances of these concepts, they become linked together by virtue of the formal coherence of the model. In the other mode of application, physical objects and the points, lines, and planes of physical geometry become linked together in the same way.

Although the logical details of the method of extensive abstraction are rather sophisticated, the overall strategy is quite simple. For example, as concerns apparent space, Whitehead begins with the idea of the relationships between any perceiver and any perceived object; then he identifies the group of apparent objects as the class containing the converse domains of these relationships. (The converse domain of the relation "father of," for example, is the class of fathers.) Then he arranges these apparent objects in converging series wherein each contains a smaller one. For example, the apparent object that is "the house" contains the apparent object that is "the room." "The room" contains "the cabinet," "the cabinet" contains "the bottle," and so on. In the end this series is perceived (or conceived) to terminate in a point or in some other basic element of perceptual geometry.

RTS is interesting in regard to Whitehead's philosophy of time for at least three reasons. First, it shows that, even as late as 1914, Whitehead was unacquainted with Einstein's work (or had just become acquainted with it and had not had sufficient time to study it). As a result, he continued to treat space and time as separate phenomena. Second, the treatise represents the first statement of an idea developed in greater detail later on: the world given in sense perception is fragmentary and imprecise and therefore quite different from the neat, orderly world of geoemtry and science. Since Whitehead had, by 1914, applied the method of extensive abstraction to time (even though he had not published it), he had by that time given up on the idea that instants of time are included in the data immediately given to sensory awareness. Rather, what is given are *durations;* that is, slices of time having temporal thickness. By applying the method of extensive abstraction to inclusion classes of durations, Whitehead showed, in the same way that he had with space, how a bridge could be constructed between the durations of sensory awareness and the temporal instants of conceptual awareness.

The third reason RTS is interesting in regard to time is that it includes a clear statement of the logic underlying the epochal theory of becoming developed in *Process and Reality*. This statement occupies practically all of Section II of the treatise and is only indirectly related to the rest of the work. Section II is devoted to an analysis of the causal relations between physical objects. Whitehead begins by identifying three axioms that govern the traditional thought on this subject: (1) one object cannot be in two places at the same time, (2) two objects cannot be in the same place at the same time, and (3) two objects at a distance cannot act on one another. Taken together, these three axioms render impossible any direct causal action between bodies as they are normally conceived. This conclusion follows quite simply: if two objects are in different places, they are at a distance from each other, and hence neither can act on the other. But if two objects are in the same place, they are the same body, and hence, once again, no action is possible. The counterargument, that action is possible between two bodies that touch one another, is easily dismissed: the notion of two contiguous physical objects is as meaningless as that of two contiguous points on a line segment.

As a solution to this problem, Whitehead suggests a structure for the physical universe according to which causal action occurs between atomic units. These units are supposed to be such that some have determinate surfaces while others do not, and those with surfaces are uniformly intermingled with those without. Given such a structure for the physical universe, the problem of contiguous physical objects does not arise, in the same way that it does not arise in mathematics for open and closed intervals uniformly interspersed on a line segment. When Whitehead adds to this the suggestion that causal action occurs not in the spatial dimension, but only in the temporal (RTS 37), the basic theory allowing causal transmission to take place between physical bodies is complete. In RTS, however, this structure for the physical universe is suggested as a mere speculative possibility, and it plays no essential role in subsequent sections of the treatise.

In the philosophy of organism, causal transmission in the form of simple physical feeling occurs exclusively in the temporal dimension. Furthermore, it occurs between one atomic entity in its phase of satisfaction and another in its initial phase of becoming. The entity in its phase of satisfaction is an entity with a determinate surface, whereas the entity in its initial phase of becoming has no surface. Thus the problem that would otherwise have arisen with contiguous entities is avoided, and it is avoided in terms of the very same suggestions about extensive relations between physical objects that Whitehead first expressed in *The Relational Theory of Space.*

Whitehead's first acknowledgment of Einstein's special theory of relativity came in an address entitled "Space, Time, and Relativity" delivered to the British Association for the Advancement of Science in 1915. From that time onward, Whitehead treated time as an integral component of a four-dimensional space-time manifold. The problem for model construction was then to develop a cognitive model that would link this manifold to our immediate sensory knowledge of nature. Constructing such a model was the primary objective of *An Enquiry concerning the Principles of Natural Knowledge* (PNK). The results of this effort were restated in more popular language in *The Concept of Nature* (CN).

In CN Whitehead uses the term *passage of nature* to designate the basic fact presented in sensory awareness. The passage of nature is the experience of a four-dimensional world that is continually "moving on." In characterizing the delivery of sensory awareness in this way, Whitehead says that he is in full accord with Bergson, except that where Bergson calls this fact "time," Whitehead calls it "passage" (CN 54). For Whitehead, nature exhibits both a spatial and a temporal passage, and so he reserves the word *time* for the one-dimensional component only.

No one, of course, experiences the full scope of the passage of nature. Rather, what we experience are temporally and spatially thick chunks of this passage. These Whitehead calls "events." Because events are extensive, they can include one another. Furthermore, this inclusion can occur in different ways. For example, the event that is the world through March 17, 1984, includes the event that is Los Angeles through that day, and the event that is Los Angeles through that day includes the event that is the Bonaventure Hotel through that day. Alternately, the event that is the world through March 17, 1984, includes the event that is the world through an hour of that day, which in turn includes the event that is the world through a minute of that day. Events of the latter sort, those that consist of a temporally thick slice of the whole world, are termed "durations." By applying the method of extensive abstraction to inclusion classes of this kind, the fundamental temporal components (moments) are yielded; by applying it to inclusion classes of the former kind, the fundamental spatial components (points, lines, planes, etc.) are yielded. A third kind of application yields the fundamental entities of four-dimensional space-time (event-particles, point-tracks, matrices, etc.).

What, then, is the meaning of time in these postrelativity works? The answer is that there are at least two meanings: there is time in the sense of the concrete, one-dimensional component of the passage of nature and the relatively abstract time of science. Both forms are, in a certain sense, subjective, and Whitehead is uncommitted about whether either form exists outside the mind. This subjective character of time can be seen from a close examination of his definition of nature. "Nature is nothing else than the deliverance of sense awareness" (CN 185). "Our experiences of the apparent world are nature itself" (R 62). In other words, nature is not something that is simply out there, but rather something that is given in *experience*. Nature *is* our experience of nature, and experience is that which occurs to a subject. When Whitehead says that "nature is closed to mind," he does not mean that nature exists independently of mind, but merely that it can be analyzed without reference to the mentality to which it occurs (CN 4–5). When understood in this light, the method of extensive abstraction succeeds in joining two aspects of human experience: the relatively concrete experience of the passage of nature with the relatively abstract knowledge of the refined entities needed for the conduct of science. The question whether these levels of experience pertain to a world existing outside the mind is a question of metaphysics, and Whitehead is not concerned with it at this time.

The general uncertainty that surrounds the relation between the experiencing subject and nature gives rise to yet a third kind of time. Whitehead calls it the "passage of mind":

So far as sense-awareness is concerned there is a passage of mind which is distinguishable from the passage of nature though closely allied with it. We may speculate, if we like, that this alliance of the passage of mind with the passage of nature arises from their both sharing in some ultimate character of passage which dominates all being. But this is a speculation in which we have no concern. (CN 69)

In *Process and Reality,* of course, Whitehead *is* directly concerned with establishing a connection between the experiencing subject and the world outside, and he shows there how both are involved in a single form of passage. The first step in this direction is taken in *The Principle of Relativity* with the interpretation of perceptual objects as "Aristotelian adjectives" (R 33) that serve as "controls of ingression" (UC 123). Having done this, Whitehead feels justified in abandoning in that work the distinction between the psychological time of the perceiver and time as it occurs in nature (R 66).

Whitehead's theory of time in CN and PNK was clearly influenced by Einstein's special theory of relativity. One of the most important implications of this theory is the existence of multiple time systems. Prior to relativity, it was thought that simultaneity had a single meaning for all perceivers and that if a certain event X was in the immediate present of P_1, it was also in the immediate present of P_2. Einstein's theory changed this; what is in the immediate present of one perceiver might be in the past or future of another. Whitehead invites us to consider two perceivers, one on earth, the other on Mars:

For the earth-man there is one instantaneous space which is the instantaneous present, there are the past spaces and the future spaces. But the present space of the man on Mars cuts across the present space of the man on earth. So that of the event-particles which the earth-man thinks of as happening now in the present, the man on Mars thinks that some are already past and are ancient history, that others are in the future, and others are in the immediate present. (CN 177)

This situation might be diagramed as in figure 5.1. E represents the perceiver on earth and M the perceiver on Mars. S_1 represents the class of event-particles in the instantaneous present of M, and S_2 the class of event-particles in the instantaneous present of E. Event-particles e_1, e_2, and e_3 are all in the earth perceiver's instantaneous present, but e_1 is in M's past, e_2 is in M's present, and e_3 is in M's future.

Problems arise from considerations such as this. Anything in the immediate present of a perceiver exists for that perceiver. Thus, event-particles e_1, e_2, and e_3 must fully exist because they are all in E's immediate present. This means that event-particles in M's future fully exist and event-particles in M's past fully exist. The same thing, of course, can be said for E in relation to M. The result is that all event-particles in the past, present, and future of all perceivers fully exist, and we seem to be left with the "block universe" advocated by relativity theorists such as Adolf Gründbaum. In a universe of this sort, all events are determined, and real becoming is impossible. Such a notion causes obvious problems for Whitehead's idea of the creative advance of nature. Whitehead does not address this problem in PNK or CN, and the cognitive model developed therein seems all but incapable of solving it. For these reasons, we will defer further consideration to the discussion of time in Whitehead's later works.

Although Whitehead accepted most of Einstein's special theory, there were two issues on which he disagreed. These concern the meaning of simultaneity and the meaning of congruence. A good deal has been written about this disagreement, which need not be recounted here.

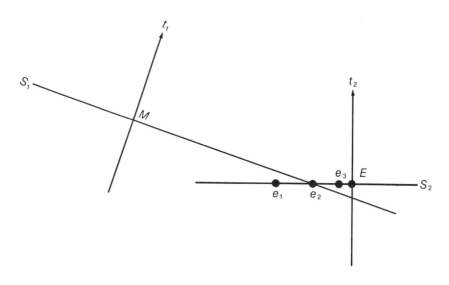

Figure 5.1.

Suffice it to note that for Whitehead, simultaneity relates to the direct meaning in the mind of the perceiver when that person thinks about an event in the immediate present. Whitehead uses this sense of simultaneity to qualify the meaning of "duration." A duration is "the whole simultaneous occurrence of nature which is now for sense awareness" (CN 53). For Einstein, on the other hand, simultaneity is a factor dependent on the transmission of light rays. Two events at a distance from each other are simultaneous if light rays coincident with their occurrence meet at a point halfway distant between them.

The disagreement about congruence involves similar issues. With regard to time, for example, how do we know that a stretch of time today—say, one minute—is the same as it was yesterday? Einstein's answer appeals to clocks, and since clocks run differently in different time systems, there is no absolute standard. Whitehead, on the other hand, appeals to the basic issue of recognition (CN 124). Recognition, he says, is an immediate factor given in sense awareness and takes place in the immediate present without the intervention of memory. In other words, we simply know when something is the same as something else, and this basic fact is required before we can begin to use clocks to measure time.

5.3. Time in Whitehead's Later Writings

Whitehead's final position on time was initiated in *Science and the Modern World* and completed in *Process and Reality* (PR). The 1926 essay entitled "Time" represents an intermediate stage of development. In PR the objective was once again to develop a cognitive model; but, unlike the earlier models, this later model is put to a unique dual purpose. First, as in the earlier writings, the model in PR is used to interpret the content of the world disclosed in sense experience; but unlike the earlier writings, this model is also used to interpret the fact of experience itself by connecting the more refined levels of experience with the more primitive, and the more primitive levels with the world outside. By proceeding in this way, Whitehead remains faithful to what he calls the "reformed subjectivist principle." According to this principle, philosophy should begin with an analysis of the content of subjectivity, but it should also show how subjectivity is linked to objectivity. This latter stage represents the "reformed" part of the subjectivist principle.

The axioms of the cognitive model developed in PR are set up in the section entitled the "Categoreal Scheme." In the briefest of terms, the dual purpose is accomplished by first interpreting the world disclosed in sense awareness in terms of actual occasions of experience and by then interpreting the experiencing subject itself in terms of the same actual occasions. In this way, both the experiencing subject itself and the world disclosed to the subject are rationalized in terms of a single cognitive model. But there is another way that the model in PR differs from the ones in the earlier works. In PR the model is a distinctively metaphysical one. Although tentative, it purports to show how the subject grows out of the world and how, within the world, the future grows out of the past. In PNK and CN, on the other hand, the model was a merely logical one dependent on the method of class constructionism, and it was used to show how the different aspects of nature might possibly be conceived as linked together. There was no attempt in those earlier works to show how they really are linked. Similar remarks extend to the model developed in MC.

Whitehead's later philosophy of time is wrapped up with this dual application of the method of model construction. Accordingly, we will begin by showing first how it relates to the interpretation of the fact of experience itself and, second, how it relates to the interpretation of the world disclosed in experience.

If we survey the content of subjective awareness, we find different kinds or levels of experience. The primitive kind is vague, heavy with emotion, and filled with feelings of derivation from the past. Whitehead's description of it sounds almost like it could have been written by Bergson:

In the dark there are vague presences, doubtfully feared; in the silence, the irresistible causal efficacy of nature presses itself upon us; in the vagueness of the low hum of insects in an August woodland, the inflow into ourselves of feelings from enveloping nature overwhelms us; in the dim consciousness of half-sleep, the presentations of sense fade away, and we are left with the vague feelings of influence from vague things around us. It is quite untrue that the feelings of various types of influences are dependent upon the familiarity of well-marked sensa in immediate presentment. Every way of omitting the sensa still leaves us a prey to vague feelings of influence. Such feelings, divorced from immediate sensa, are pleasant, or unpleasant, according to mood; but they are always vague as to spatial and temporal definition, though their explicit dominance in experience may be heightened in the absence of sensa. (PR 267)

The kind of experience Whitehead describes here provides our most basic acquaintance with time. Rather than call it "time in the primitive sense," however, Whitehead calls it "becoming." Becoming is experienced as the concrete awareness of the endurance of the self. It is the experience of the influx into ourselves of the world outside and of the internal transition of feelings one into another. As Whitehead observes, the awareness of this endurance is most prominent in the *absence* of sensa. It is quite different from the sort of time involved in the awareness of external events, such as the motion of a car or the falling of a stone.

Whitehead accounts for this primitive sense of time in terms of the theory of causal efficacy. In general, causal efficacy is a mode of perception that is present whenever there is a feeling of derivation or causal influence. The most basic form is provided by simple physical feelings; that is, feelings by which an occasion prehends a feeling in the satisfaction of an antecedent occasion. Because the subjective form of a simple physical feeling conforms to the subjective form of the feeling felt, there is a flow of causal influence from the antecedent occasion to the immediate one. But causal efficacy is also involved in the purely intrasubjective process by which the eternal objects ingredient in simple physical feelings are selected and reinforced through the category of transmutation (PR 478). In this latter capacity the self is aware of its own efficacy as a constructive activity.

The second important level of experience that Whitehead accounts for in terms of the categoreal model of PR is ordinary sense perception. This is the level of experience familiar to all of us and the one that modern philosophy attended to so exclusively. It consists of the awareness of clearly demarcated sensa projected against a spatiotemporal backdrop. In this level of awareness time appears as a one-dimensional component in the four-dimensional backdrop against which the sensa are projected. Whitehead's systematic term for this level of awareness is perception in the mode of presentational immediacy:

> Presentational immediacy is an outgrowth from the complex datum implanted by causal efficacy. But by the originative power of the supplemental phase, what was vague, ill defined, and hardly relevant in causal efficacy, becomes distinct, well defined, and importantly relevant in presentational immediacy. (PR 262)

By the "originative power of the supplemental phase," Whitehead means the process by which eternal objects ingredient in the physical pole are prehended by the mental pole through conceptual reproduction and then reintegrated with the physical pole through transmutation. The selective reintegration results in the reinforcement of certain features of the simple physical feelings prehended in causal efficacy. These features as reinforced are the sensa that appear in presentational immediacy. The spatiotemporal backdrop against which these sensa are projected is provided by a different kind of feeling, which Whitehead calls "strain feelings." The model for strain feelings was provided by the stresses and strains that Maxwell visualized as distorting the electromagnetic field. These strain feelings, Whitehead says, allow us to discriminate regions not only in space but also in time and, in this way, to determine exactly "where" a perceptual object is. Time, as it functions in presentational immediacy, is a one-dimensional component in the four-dimensional datum provided by strain feelings.

The third level of experience that is important in connection with time is conceptual awareness. Not only do we perceive the passage of time through visual and auditory sensations, but we know what time is conceptually. We have a concept of what a duration is, what a moment is, how moments are serially ordered within a duration, and how they are featured within the four-dimensional geometry of the world. This is time in the abstract, the time that appears in the mathematical equations of physics and chemistry.

Whitehead accounts for abstract time in terms of intellectual feelings. An intellectual feeling is a highly complex kind of comparative feeling that is normally derived from presentational immediacy and that has consciousness as its subjective form. Abstract, conceptual time may be interpreted as either the datum of such an intellectual feeling or as the datum of a conceptual feeling that reproduces the eternal object ingredient in an intellectual feeling.

Thus far we have considered time as it occurs in the perceiver/thinker. The other major occurrence of time is in the world existing outside the perceiver. The interpretation of the content of the external world is the second aim of the model developed in PR. This content includes trees, mountains, galaxies, symphony concerts, other perceivers, and so on. The fundamental relational network underlying this vast, four-dimensional universe of entities is the extensive continuum.

The extensive continuum is that general relational element in experience whereby the actual entities experienced, and that unit experience itself, are united in the solidarity of one common world. The actual entities atomize it, and thereby make real what was antecedently merely potential. The atomization of the extensive continuum is also its temporalization; that is to say, it is the process of the becoming of actuality into what in itself is merely potential. (PR 112)

Prior to its actualization through ingression in actual occasions, the extensive continuum is only potential. This means that, prior to actualization, the potential extensive continuum is either an eternal object or collection of eternal objects. As such, it is not even extensive—just as the concept of triangularity is not itself triangular. When the extensive continuum is actualized, extensiveness makes its appearance. Furthermore, Whitehead says, the actualization (atomization) of the extensive continuum is its "temporalization." Whitehead could have added the statement that the actualization of the extensive continuum is also its spatialization. Both space and time involve dimensionality, and the bare extensive continuum involves no dimensionality. The temporalization and spatialization of the continuum occurs through the welding of dimensionality with extensiveness in the concrescence of an occasion. At the same time, of course, what was previously only potential has become actual. Thus, the world external to the perceiver can be said to exhibit two kinds of time: potential time, which consists in the potential unification of the two potentials that are extensiveness and dimensionality, and actual time, which is a feature in the satisfaction of an actual occasion.

A point that Whitehead stresses several times in PR is the idea that "extensiveness becomes, but 'becoming' is not itself extensive" (PR 53):

The conclusion is that in every act of becoming there is the becoming of something with temporal extension; but that the act itself is not extensive, in the sense that it is divisible into earlier and later acts of becoming which correspond to the extensive divisibility of what has become. (PR 107)

In other words, it is the final phase of satisfaction, and only that phase, that is extensive. Whitehead reaches this conclusion from an analysis of Zeno's arguments. If becoming were extensive, there would be an earlier half and a later half of every concrescence. The earlier half, in turn, could be divided into an earlier quarter and a later quarter, and so on indefinitely. Since the earlier of each of these divisions must occur before the one that succeeds it, the conclusion is that the concrescence can never get started (PR 106). Whitehead solves the problem by taking becoming outside of extensiveness altogether. Accordingly, there is no earlier half and no later half. Becoming is like a fifth dimension in Whitehead's philosophy—if there is such a thing as a dimension that is not extensive. When, at the end of the concrescence, the occasion reaches satisfaction, a four-dimensional quantum appears in the world objectified in presentational immediacy. In respect to time, this quantum expresses the fundamental insight of the "epochal theory of time" (PR 105).

When an occasion attains satisfaction, it becomes an object for subsequent occasions and is susceptable of being analyzed. Two kinds of analysis are possible: genetic analysis and coordinate analysis. Genetic analysis discloses how the phases of concrescence grew out of one another. In other words, it discloses the relationships between the various feelings in the concrescence and eventuates in the theory of feeling in PR. Coordinate analysis, on the other hand, discloses the spatiotemporalness of the occasion. "Physical time makes its appearance in the 'coordinate' analysis of the 'satisfaction.' The actual entity is the enjoyment of a certain quantum of physical time" (PR 434). This quantum is continuous and susceptible of infinite divisibility. Thus, it provides the ultimate external referent for time as it appears in the formulae of the physical sciences.

By way of summary, Whitehead's later philosophy includes five basically different (but interconnected) kinds of time:

1. time in the sense of becoming,

2. sensory time,

3. abstract, conceptual time,

4. potential physical time, and

5. actual physical time.

Time in the sense of becoming exists in all actual occasions and is disclosed in feelings of causal efficacy. Sensory time is the time disclosed in feelings of presentational immediacy and is present only in those occasions that are sufficiently sophisticated to have such feelings. Abstract, conceptual time is derived from intellectual feelings and is found only in high-grade occasions. Potential physical time is an aspect of the extensive continuum prior to its actualization by occasions, and actual physical time is a feature of the satisfaction of all actual occasions. The relationship between time in the sense of becoming and actual physical time may be illustrated in figure 5.2.

The shaded portion represents the spatiotemporal quantum exhibited in the satisfaction of an occasion, with its *s* and *t* axes indicating the dimensions of physical space and physical time. The *b* axis represents the dimension of becoming, through which the occasion concresces and eventually attains satisfaction, and *I* represents the point of initial aim. The initial aim is properly represented as a point because the occasion at this stage (and, indeed, at all stages leading up to the satisfaction) is not extensive. Extensiveness (the shaded part) becomes, but becoming is not itself extensive. The trouble with this diagram is that it *appears* to represent the dimension of becoming not only as extensive but as perpendicular to the dimensions both of physical time and physical

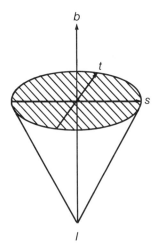

Figure 5.2.

space. Properly conceived, becoming is neither extensive nor perpendicular to any other dimension.

No one, of course, can *see* an actual occasion in the external world. What we see are nexūs of occasions. A nexūs in which a common element of form is causally transmitted from earlier to later members is termed a "society." Time makes its appearance as a feature of societies; namely, the feature by which societies endure. Whitehead's theory of society is a study in itself, and I cannot go into it here. I can note, however, that the kind of society that grounds our perception of time is what Whitehead calls a corpuscular society. An example would be a stone or a tree or any other sense object that endures through time. A corpuscular society is composed of strands of personally-ordered societies; that is, groups (or nexūs) of serially-ordered occasions that exhibit the transmission of a common element of form. Corpuscular societies display visible regions of the extensive continuum that arise from the synthesis of the quanta featured in the occasions that make up the society.

Whereas every society is a nexus, not every nexus is a society. Specifically, nexūs in which the members are causally independent are nonsocial nexūs. Every occasion has such a nexus associated with it, and the members are called the "contemporaries" of the occasion (PR 192). In other words, the contemporaries of an occasion M are the occasions that are neither in the causal past nor the causal future of M. This notion may be illustrated as in figure 5.3.

Where s and t designate the space and time axes, respectively, and L_1 and L_2 the paths of the fastest causal transmissions (i.e., light rays), the region below M consists of the occasions that provide data for M, and the region above M consists of the occasions for which M will provide data. The regions on either side of M include the occasions causally disconnected from M. These are M's contemporaries.

A duration is any temporally thick slice of space-time within the nexus of M's contemporaries that is characterized by the fact that any two members of the slice are contemporaries. All the members of a duration are said to be in "unison of becoming" (PR 192). Any occasion M lies in many durations, but there is one duration that is special in that it contains the occasions that M perceives in the mode of presentational immediacy. This duration is called M's "presented duration" (PR 192). Where D_1 and D_2 represent durations including M, and D_p represents M's presented duration, this idea may be diagrammed as in figure 5.4.

As the diagram illustrates, Whitehead, in his later writings, defined "duration" differently from the way he defined it in the early works.

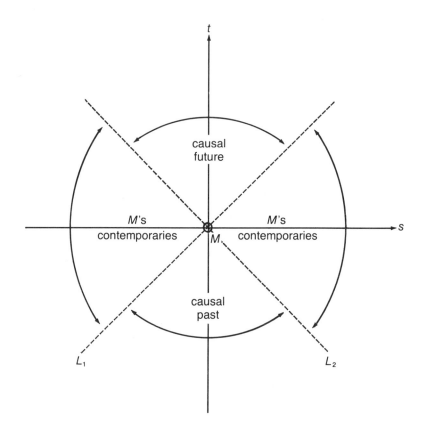

Figure 5.3.

In the early works, only one of the slices contained in the region of *M*'s contemporaries was called a "duration." This was the slice that included the world disclosed in sense perception. In the later writings, as we have noted, this special slice is called the "presented duration."

Let us now return to the "block universe" problem discussed in the previous section. Suppose that *M* and *N* define different inertial frameworks such that *N* lies in *M*'s presented duration, but *M* does not lie in *N*'s presented duration. Instead, *P*, which is in *M*'s causal past, lies in *N*'s presented duration (see figure 5.5).

The problem is that *N* is in unison of becoming with *M*, and *P* is in unison of becoming with *N*, but *P* is not in unison of becoming with *M*. *P* is in *M*'s causal past and has therefore perished for *M*. The

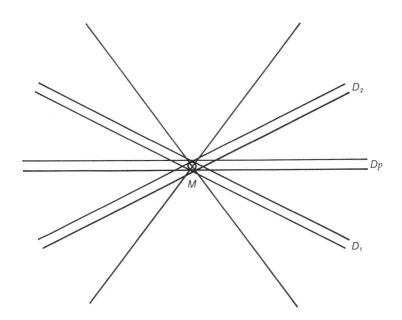

Figure 5.4.

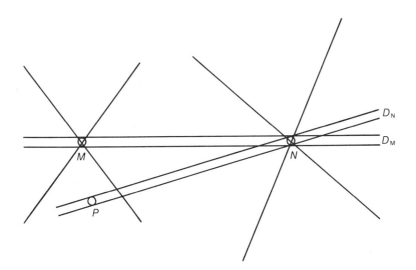

Figure 5.5.

difficulty could be avoided, of course, by denying the transitivity of unison of becoming, but this would seem to violate Whitehead's intentions. Unison of becoming is an existential relation that seems to be more fundamental than the mechanics of time systems.

A possible solution to this difficulty may lie in the observation that in Whitehead's later theory of time, causal transmission occurs in the "dimension" of becoming—not the dimension of physical time. Thus, the letter P in the diagram does not stand for an *occasion* in the causal past of M, but merely for a spatiotemporal quantum that was once "occupied" by such an occasion. This quantum is now "occupied" by a new occasion in concrescent unison with N. The occasion that was in M's causal past, and that transmitted its data to M in the dimension of becoming, has indeed perished and has dropped out of the picture. This solution, if it be such, requires a clarification of the terms *causal past* and *causal future*. Here *past* and *future* refer to becoming—not to physical time. The causal future of M consists in the occasions that will become out of the data transmitted by M, and the causal past of M consists in the occasions that are now reenacted in M's becoming.

It is interesting to compare this solution with the one Bergson would appear to offer. In *Time and Free Will*, Bergson argues persistently against what he calls the "spatialization of time"—the representation of time in diagrams such as the ones presented above. These sorts of representations, he contends, misrepresent the true character of time. Whitehead, on the other hand, defends this practice:

Thus in so far as Bergson ascribes the "spatialization" of the world to a distortion introduced by the intellect, he is in error. This spatialization is a real factor in the physical constitution of every actual occasion belonging to the life history of an enduring physical object. (PR 489)

For Whitehead, the spatialization of time is required for the normal conduct of science, but it should be noted that this spatialization attaches only to physical time and abstract time. The one kind of time that is definitely not spatialized in Whitehead's later philosophy is becoming, and this is the sense of time that is closest to Bergson's nonspatialized time. Thus, in this respect, Bergson's and Whitehead's positions are not so very far apart.

Abbreviations of Whitehead's Works

AB Notes "Autobiographical Notes," in *Science and Philosophy* (New York: Philosophical Library, 1948)

CN   *The Concept of Nature* (Cambridge: Cambridge University
        Press, 1907)
MC   *On Mathematical Concepts of the Material World,* in *Alfred
        North Whitehead, An Anthology,* ed. F. S. C. Northrop and
        Mason W. Gross (New York: Macmillan, 1961)
PNK   *An Enquiry concerning the Principles of Natural Knowledge*
        (Cambridge: Cambridge University Press, 1919)
PR   *Process and Reality: An Essay in Cosmology* (New York:
        Harper and Row, 1960 [1929])
R   *The Principle of Relativity with Applications to Physical
        Science* (Cambridge: Cambridge University Press, 1922)
RTS   "The Relational Theory of Space," in "Whitehead's 'Re-
        lational Theory of Space'—Text, Translation, and Com-
        mentary," by Patrick J. Hurley, *Philosophy Research Ar-
        chives,* vol. 5 (1979)
SMW   *Science and the Modern World* (New York: Macmillan,
        1925)
UC   "Uniformity and Contingency," in *The Interpretation of
        Science,* ed. A. H. Johnson (New York: Bobbs-Merrill, 1961)

Notes

1. Silvanus P. Thompson, *The Life of William Thomson, Baron Kelvin of Largs,* 2 vols. (London: Macmillan, 1919), 2.830.
2. George Gabriel Stokes, *Memoir and Scientific Correspondence,* ed. Joseph Larmor, 2 vols. (Cambridge: Cambridge University Press, 1907), 1.218.
3. William Christian argues for a similar linguistic division. See his *Interpretation of Whitehead's Metaphysics* (New Haven: Yale University Press, 1967), 3.
4. Bertrand Russell, *The Autobiography of Bertrand Russell,* 3 vols. (Boston: Little, Brown, 1951), 1.188. The same statement occurs in his *Portraits from Memory* (New York: Simon and Schuster, 1951), 100.
5. See the comment about Whitehead's treatment of instants of time in Russell's *My Philosophical Development* (New York: Simon and Schuster, 1959), 103.

6. Frederick Ferré

CONTEMPORANEITY, KNOWLEDGE, AND GOD: A COMMENT ON HURLEY'S PAPER

The purpose of this comment is to focus on three significant issues that can be drawn from the well-laid feast of ideas put before us by Patrick Hurley. I propose that we first discuss Whiteheadian epistemology in view of his theory of time, then note the challenging implications of these matters for Whitehead's theology, and finally reflect on the logic of models in Whiteheadian theorizing.

(1) The epistemological mode of causal efficacy poses no special problems, once the general Whiteheadian scheme of prehension and concrescence is accepted. (Causal efficacy, after all, is a direct exemplification of the most basic categories of causal transmission between entities.) The notion of causal efficacy is in my opinion the most important of Whitehead's epistemological contributions, not only because this dimension of perception has been, especially among English-speaking philosophers, the most neglected cognitive phenomenon, but also because it is the basis of Whitehead's justification of induction and thus of his reply both to Hume and to Kant regarding reliable knowledge of the world. Although Whitehead's discussion of causal efficacy was phenomenologically quite revolutionary, anticipating by many years Continental discussions of "embodiment," and the like, it was scientifically entirely orthodox. David Bohm expresses this well in his chapter, "Time, the Implicate Order, and Pre-Space":

In general, what we are conscious of as now is already past, even if only by a fraction of a second. . . . Therefore, from the past of the present, we may be able to predict, at most, the past of the future. The actual immediate present is always the unknown.

Whitehead, of course, agrees that contemporary occasions, i.e., actual occasions in concrescent unison with one another, cannot be ingredient in each other's causal histories and therefore cannot be known through the mode of causal efficacy. Bohm's logical inference from this bit of physics and physiology to the conclusion that "the actual immediate present is always the unknown," however, seems contradicted by Whitehead's epistemological mode of presentational immediacy. In this mode, if it exists, contemporary occasions can be cognizant of one another. As Hurley puts it:

In the early works, only one of the slices [of the region of an occasion's contemporaries] was called a "duration." This was the slice that included the world disclosed in sense perception. In the later writings, . . . this special slice is called [the occasion's] "presented duration."

Whitehead himself wants to preserve a cognitive relationship between "the 'percipient at that moment,' and that other equally actual entity, or set of entities, which we call 'the wall at that moment.' "[1] He speaks of presentational immediacy as "the familiar presentation of the contemporary world."[2]

To be sure, the relationship is one based on "projection" onto the contemporary spatial region, by the percipient, of qualities concretely received through the mode of causal efficacy. This "projection" is not a mere addendum, however, since it is "an integral part of the situation, quite as original as the sense-data."[3] It means that the color of the wall, as an abstract entity now "part of the make-up of the percipient," is "at that moment indifferent to the wall."[4] In this way Whitehead hopes to preserve his principle that "contemporary events happen independently"[5] together with the traditional, prescientific view that the "familiar immediate presentation of the contemporary world" can in fact be veridical.

Whitehead, however, can hardly have it both ways. Presentational immediacy is experienced phenomenologically as not merely inferential (it is what some philosophers would call the basis of our primary experience of intentionality), but inasmuch as it is fundamentally "an outgrowth from the complex datum implanted by causal efficacy,"[6] it cannot, even for Whitehead, finally be a mode (literally) of independent *presentation* of the contemporary world at the moment. Rather, it is a dependent and derivative mode of *representation,* reflecting the immediate past of the percipient, just as David Bohm asserts. Contemporary occasions, strictly speaking, if they truly "happen independently," cannot but be ignorant of one another in that moment of independence. The greater the remove of the contemporaries from recent relevant "implantations" of data through chains of causal efficacy, the deeper the ignorance.

(2) The idea of the independence of contemporary occasions creates the much-discussed problem for Whiteheadian theology of how God, if described as the one nontemporal actual entity, can possibly know the world or be known by it. All temporal entities reach satisfaction, become objective, perish—but in perishing become data for new events of causal efficacy as fresh concrescing entities prehend these data in advancing the everlasting rhythms of the world. The one "non-tem-

poral" actual entity, however, on Whitehead's scheme, never perishes; thus God is relegated to the status of being a contemporary to every temporal actual entity.

The problem raised by Whitehead's epistemology for his theology is quite sharp: as long as God is described as external to and interactive with the world, then either God must be causally related to the world's epochal occasions or there seems to be no clear basis for his knowledge of or influence on them. God cannot know the world by presentational immediacy alone since this is not an independent mode of knowledge but is merely a way in which "contemporaries in unison of becoming" (from not-too-distant regions) may noninferentially represent to themselves those relevant regions by means of data concretely implanted—in the case of God, through hybrid physical feelings[7]—in the immediately prior moment through the mode of causal efficacy. But if God participates with the world in cognitive and causal ways, he must himself share the rhythms of temporal occasions. He must, that is, have an initial phase in which he prehends his immediate past environment (from which he gathers his data) and he must come to some epochal satisfaction. The theoretical motivation for the development of theological temporalism—despite its own problems, especially in jeopardizing the self-identity of God—is very clear.

There seem, then, as John Cobb put it two decades ago, to be "elements of incoherence in Whitehead's doctrine of God"[8] if God is to be portrayed as one who can actively participate in "luring" the world (through the Category of Reversion) toward greater intensity of value and also at all moments be the world's contemporary. A God not in touch with the temporal occasions of the world through causal efficacy would seem to be simply irrelevant, on Whitehead's own showing; but a causally efficacious God would need to be epochal in order not to be the dreaded "exception to the metaphysical categories" so repugnant to Whitehead's theoretical method and commitments. The main point, however, is not merely that Whitehead's scheme needs improving more than Hurley suggests but, still more important, that Whitehead's methods continue to be applied in vital theological construction and that Whitehead's philosophic commitments continue to inspire his successors.

(3) Whitehead's theoretical commitments, as Hurley shows us, made him deeply intent on manifesting connections and coherences among the many apparent discontinuities of thought and experience. I am not sure how helpful it is to refer to Whitehead's work as "modeling"

activities, especially if the paradigm of modeling is taken initially from Lord Kelvin's mechanical literalism. Kelvin declared that nothing deserved to be considered understood or understandable unless a mechanical model could be built to illustrate it. The notion of what Hurley calls a "cognitive model" was far from Kelvin's goal, and it would be an impossible task to satisfy Kelvin today, either in metaphysics (can one imagine building a mechanical model to show how the phases of concrescence work?) or even in his own wonderfully transformed field of physics.

Whitehead, I think, was "modeling" only in the weak sense that any theorizing can be said to model a subject matter; but then the useful distinction between model and theory is lost.[9] Much more work needs to be done in interpreting with detailed and heuristically fruitful models the abstract formalisms of Whitehead's thought. Indeed, one of the urgent tasks for defenders and extenders of Whiteheadian theory lies precisely in continuing to identify and explore vivid imaginative images or metaphors by which they and others may clothe this powerful scheme of ideas.

Be that as it may, Hurley is quite right in stressing the Whiteheadian aim at connecting "disparate classes of phenomena." This aim is what makes continued work within Whiteheadian theory so valuable, since it so largely succeeds in showing one way to get beyond the "bifurcations" that plague modern thought—and life. Biology and religion, physics and poetry, mathematics and history, all need the synoptic vision that his speculative categories boldly offer. This integrative aim is also what makes the criticism and correction of Whitehead's theories worth our while, including the addressing of incoherences such as those that rise from his inclination to remove his God from the perturbations of temporality[10] and at the same time to insist that his God must be the supreme exemplification of the metaphysical categories.

Both these inclinations are understandable and valuable. To drop the latter would be to give up the struggle for comprehensive coherence; to drop the former would be to let go of the deep intuition found in much of humanity's religious history: that time when felt merely as "gnawing tooth" is somehow not the full story. As we search for a satisfying solution—for a wider coherence within a fuller adequacy— at least we may know that we remain within an ongoing enterprise of conceptual construction supported, as Whitehead wrote, by "that counter-tendency which converts the decay of one order into the birth of its successor."[11]

Notes

1. A. N. Whitehead, *Symbolism, Its Meaning and Effect* (Macmillan, 1927), 15.
2. Whitehead, *Symbolism,* 13.
3. Whitehead, *Symbolism,* 14.
4. Whitehead, *Symbolism,* 16.
5. Whitehead, *Symbolism,* 16.
6. A. N. Whitehead, *Process and Reality* (Macmillan, 1929), 262.
7. Whitehead, *Process and Reality,* 163, 343, 375–77, 469.
8. John B. Cobb, *A Christian Natural Theology* (Westminster Press, 1965), 176–77.
9. Frederick Ferré, "Mapping the Logic of Models in Science and Theology," *Christian Scholar* 46, 1 (Spring 1963).
10. For a thoughtful defense, however, of Whitehead's inclination in this direction, see Bowman L. Clarke, "God and Time in Whitehead," *Journal of the American Academy of Religion* 48, 4 (1980).
11. A. N. Whitehead, *The Function of Reason* (Beacon Press, 1929), 90.

7. Peter Miller

TIME, EVENTS, AND SUBSTANCE:
COMMENTS ON HURLEY AND WHITEHEAD

Let me say first how much I appreciated Patrick Hurley's paper. It has certainly given me a more complete perspective on Whitehead's thought and helped me to focus on some of the problems of thinking about time. Although like others at this conference I have found inspiration and illumination in Whitehead's philosophy, my approach to metaphysics differs from his. In a nutshell, I am an Aristotelian who thinks that enduring individual substances are ontologically more fundamental than events, in that the latter are features of the former rather than vice versa. Thus my comments will address some points in Hurley's interpretation but additionally will raise questions regarding the tenability of portions of Whitehead's philosophy. I shall address in turn problems raised by relativity theory, epochalism, and the mathematization of reality.

7.1. Relativity and the "Block Universe"

Hurley raises the problem of whether relativity theory might entail a "block universe" that precludes objective becoming. In the "block universe," past, present, and future are timeless (eternally real) relational attributes between events. Although an event P might timelessly be past relative to one event and present relative to another, it could not timelessly (or at any time) bear the two nonrelational and mutually exclusive ontological attributes of "becoming" and "having perished." How then could event P be in unison of becoming with N, which is in unison of becoming with M, while also lying in M's causal past and thus having perished for M?[1]

Hurley's interpretation of Whitehead's solution to this dilemma is to say that when events M and N are co-present (i.e., in unison of becoming), the event that at one time occupied spatiotemporal quantum P is dead and gone but replaced by another event at P in concrescent unison with N (and thus presumably also with M).

This interpretation, I believe, will not wash. In the first place, Whitehead's actual occasions seem to be cemented each into its own unduplicated standpoint within the continuum. Even if two could occupy the same quantum, the problem would remain because the continuum (as atomically actualized) reflects the causal order among occasions.[2] Any occasion at P would presumably belong to the past actual world of M while being in unison of becoming with N. Whitehead says that in the mode of presentational immediacy we draw on data from the past to "illustrate" *potential* regions of the space-time continuum displayed before us, but the one and only *actual* occasion out there at P is really in concrescent unison with N and consequently not discernible in its concreteness by N.[3] So the problem remains: how can both P and M be in unison of becoming with N without having earlier and later events becoming simultaneously?

Whitehead's explicit answer, I think, is precisely the one Hurley dismisses as contrary to Whitehead's intentions: he affirms the multiplicity of sets of events in unison of becoming with a given event ("durations") and the relativity of simultaneity (as defined by "presented durations") and thus denies the transitivity of "unison of becoming." Hence there is no single absolute, unambiguous, cosmic well-ordered sequence of becoming among actual occasions that are causally unrelated.

Many Whiteheadians, beginning perhaps with Whitehead himself, who denied the metaphysical necessity of relativity, have shared Hurley's discomfort with this implication. Is this discomfort just a case of prerelativistic habits of thought dying hard, or is there a more fundamental ground? Specifically, is there anything in the metaphysical notion of objective becoming or in physical theory or in other well-founded metaphysical principles that does not cohere well with the indeterminate (or multiple and relativistic) forms of synchronization of the becoming of remote events?

Hurley thinks that "unison of becoming is an existential relation that seems to be more fundamental than the mechanics of time systems." So it is, but (as Whitehead defines the expression) it is just the relation of mutual causal independence that does not preclude multiple durational slices that satisfy it. Nor do we have to say that a given event P either timelessly or at any particular time had the contrary attributes of "becoming" and "having perished." P became before it perished in any and every time system that the extensive continuum coordinates.[4]

Do new developments in physical theory, such as Bell's theorem in quantum mechanics, lead to a nonrelativized interpretation of "unison of becoming"? That is not altogether clear from Stapp's treatment of the subject, and most Whiteheadians who welcome this development do so, like Hartshorne, because it appeared to resolve other metaphysical difficulties.[5]

In a most general fashion I venture that the metaphysical difficulties behind Hurley's and Hartshorne's and others' desire to get beyond relativity stem from a perceived need to coordinate causally independent and spatiotemporally remote events when and as they become. The extensive continuum does this in a formal way, but some form of noncausal existential tie between contemporaries is also thought to be needed. Whitehead assigned a role like this to God, and Hartshorne apparently thinks that divine omnipresence and cosmic influence require a remote synchronization of events that would establish a preferred present duration among the multitude that relativity theory allows.[6] But whatever one makes of the merits of this metaphysical conviction, it is important to distinguish it from Whitehead's relativistically defined technical expression, *unison of becoming.*

7.2. Epochalism and the Phenomenology of Time

In the balance of these remarks I wish to take issue not with Hurley but with Whitehead himself and to sketch an Aristotelian alternative to some of his basic tenets. Hurley usefully points out that time is a multifaceted concept and that at least five aspects of time are to be found in Whitehead's thought: becoming, sensory time, conceptual time, potential physical time and actual physical time. Phenomenologically the most important of these are becoming and sensory time, because conceptual time is constructed from abstractions of sensory time primarily (at least in the physical sciences), and physical time is a metaphysical generalization of the former two to the world at large. This generalization occurs in accordance with the reformed subjectivist principle, which holds, as Hurley put it, that "philosophy should begin with an analysis of the content of subjectivity, but it should also show how subjectivity is linked to objectivity."

The phenomenon of time for Whitehead thus contains two distinguishable aspects: a chronology, which Whitehead correlates with perception in the mode of presentational immediacy, and the dynamism of becoming, perceived in the mode of causal efficacy. The latter is what is particularly characteristic of process philosophy to note and to generalize as an ultimate objective feature of the universe at large. Andrew Bjelland's paper on Bergson analyzes this becoming in terms of "durational succession" as a dynamic, asymmetrical, and internal relation between successive portions of temporal process. Bjelland points out that "this relational pattern is intrinsic both to existential-phenomenological theories of intentionality and to Whitehead's theory of prehensive concrescence." But there is an important distinction between Bergson and the phenomenologists on the one hand and Whitehead on the other. As Bergson reports our experience of time, the distinction of successive moments into discrete events is a product of analysis after the fact rather than an intrinsic characteristic of the process of becoming itself:

Whilst I was experiencing them [i.e., successive moments of the flux of experience] they were so solidly organized, so profoundly animated with a common life, that I could not have said where any one of them finished or where another commenced. *In reality no one of them begins or ends, but all extend into each other.*[7]

Whitehead, by contrast, insists that becoming exists in discrete packets as actual occasions that atomize the spatiotemporal continuum. One occasion must become, reach satisfaction, and perish before the next can come to be in any timelike sequence. In order to draw out the contrast I shall sketch briefly an alternative, Aristotelian, substantialist interpretation of the continuous dynamic, asymmetrical, and internal relations that Bergson recognized in duration.

In the Aristotelian tradition, these features of the creative process of the world are to be found in the activities and interactions of natural substances. What is a substance? The Aristotelian paradigm of a substance is an individual organism, a man or a horse, for example. These and other kinds of concrete individuals are ontologically fundamental because everything else is an aspect, property, compound, or relation of these. A substance is not an unintelligible surd or propertyless material substratum underlying the phenomena; rather, it is a concrete whole, which we can comprehend because we can both confront it in experience and abstract its formal characteristics for contemplation and analysis. Nor is a substance inert; it maintains its nature in the face of, and imposes itself upon, an ambient and often contrary world, and it may develop in accordance with its characteristic potential. And, most important for our purposes, a natural substance is essentially temporal because it possesses numerical identity through change or mutable durability. The abstractable qualities it contains have varying degrees of permanence and temporariness. It is a mistake of Hume's to think that permanence or persistence and change are mutually exclusive. In fact, permanent things *must* change, at least in temporal locus and relations to other changeable things and passing events, and change can occur only to that which endures. Kant and Aristotle were right in noting that these concepts presuppose one another. The contrary of the permanent is not the changing but the changeless, which is confined in nature to a particular condition, time, and circumstance, like Whitehead's actual occasions. What cannot change must perish.

The substantialist alternative to epochalism better accords with the phenomenology of perception and of agency, I think. A proper phenomenology of perception would note that the perception of an object is time-bound. We see an aspect or facet of an object in the moment with expectations as to how the experience will unfold as we continue to observe and interact with the object. Subsequent experience fulfills and modifies those expectations while elaborating further expectations for the more remote future. The same I who experiences an object at one moment in time finds its prior expectations confirmed and disconfirmed in various ways through subsequent experience. I, the experiencer, at least, am a substance, a mutably durable identity through time.

But I am not only an experiencing subject; I am an *agent,* responsible for my character and deeds. As such I need time for (1) growth in knowledge of the world's and my own potentialities, (2) formation of long-range goals and values, (3) awareness of present circumstances, (4) formation of specific intentions, (5) initiation of actions, (6) feedback on progress, (7) modification of the course of action, (8) evaluation of successes and failures of attempts, (9) reevaluation and re-formation of intentions, (10) reevaluation and re-formation of long-range goals and values, and so on. Thus the various aspects of my agency for which I and others hold myself responsible require that I endure for a considerable stretch of time. No temporal slice of my existence, particularly if it is below the threshold of conscious modification, can be held similarly responsible. If such a slice were self-determining, it could be so only in a nonresponsible way. How could a sequence of nonresponsible segments add up to a responsible agent?

Given this *prima facie* case for substantialism, we must ask: Does Whitehead have adequate grounds for rejecting the continuous mutable durability of temporal process in favor of the atomization of time and process by actual occasions in a manner that, on the face of it, is contrary to the phenomenological evidence? Are there considerations in mathematics, physics, or systematic philosophy that drive us to this outcome? And if there are, how can the experienced *durée* of time be reconstructed from a succession of immobile atomic occasions? Or, put self-referentially, has Whitehead violated his own reformed subjectivist principle or committed the fallacy of misplaced concreteness by reifying the products of analysis? There is much to be said by way of a Whiteheadian reply to these questions, but these few remarks will have to suffice for now to sketch the field of debate.

7.3. Mathematics and Reality

My final set of remarks are, I'm afraid, much more inchoate than the previous ones, but their gist is a complaint that Whitehead is Pythagorean in his mathematization of reality. I believe Whitehead once admonished that we should seek precision and then distrust it. One of the fascinations of *Process and Reality* is to see his attempts to give precision to his ideas, but I distrust some of the results.

Again Patrick Hurley has illumined this point for me in his discussion of Whitehead's early work, *The Relational Theory of Space,* in which Whitehead struggles with the question of what sort of structure is presupposed if physical existents are to have causal interfaces. Whitehead concludes that alternating determinate and indeterminate surfaces of physical objects, analogous to interspersed closed and open intervals on line, are needed. Hurley interprets Whitehead's mature philosophy to retain this idea in the structure of each actual occasion: "The entity in its phase of satisfaction is an entity with a determinate surface, whereas the entity in its initial phase of becoming has no surface." If we understand this on the analogy of open and closed intervals, we could interpret it to mean that pastwards an actual occasion in its immediacy begins at the surface of its immediate predecessors, which, on the assumption of classical relativity theory, must also be in immediate spatial proximity, whereas present- and futurewards the occasion contains its own boundaries and thus is causally insulated from all similarly self-contained contemporaries, even though these may be adjacent to it.

Note that an open interval is no less determinate than a closed interval. Each has perfectly precise limits, the only difference being that in the one case the limits are internal to the interval and in the other, external. An actual occasion thus has a regional standpoint in the extensive continuum, which is perfectly precise on all sides. As Whitehead says:

The inside of a region, its volume, has a complete boundedness denied to the extensive potentiality external to it. The boundedness applies both to the spatial and temporal aspects of extension. Wherever there is ambiguity as to the contrast of boundedness between inside and outside, there is no proper region.[8]

I have several difficulties with such a picture of actuality:

1. It puts God in a deus-ex-machina role of assigning standpoints and initial aims to concrescing occasions. This is necessary in order to provide continuity of pursuits from one

occasion to the next and to harmonize standpoints so that they together constitute a plenum without a shade of overlap or interaction. This implicates God in, say, the continuation of the cancer that kills the child and thus raises severe problems for theodicy.

2. In spite of Whitehead's great stress on the decisionmaking powers of an actuality, it greatly restricts or denies these powers with respect to shape, location, or boundary and makes metaphysically impossible jointly or interactively determined boundaries.

3. I have already objected to the sharp temporal boundaries of atomic occasions as contrary to the continuous *durée* of phenomenological time.

For these reasons, I think Whitehead is guilty of what I call "the fallacy of precise location" (not to be confused with simple location)—confusing idealized points and boundaries without thickness with the thick spatiotemporal boundaries and divisions of the actual world.[9]

Notes

1. See figure 5.5 and accompanying text in Hurley's chapter.
2. See my comments in "On 'Becoming' as a Fifth Dimension," below.
3. *Process and Reality,* part 2, chaps. 2 and 4.
4. See M. Čapek's chapter, below.
5. See C. Hartshorne, "Bell's Theorem and Stapp's Revised View of Space and Time," *Process Studies* 7, 3 (1977):183–91.
6. Hartshorne, "Bell's Theorem," 187.
7. Henri Bergson, *An Introduction to Metaphysics* (1913), as cited in *Problems of Space and Time,* ed. J. J. C. Smart, 139–40; emphasis added.
8. *Process and Reality,* Corrected edition, ed. D. R. Griffin and D. W. Sherburne (New York: Free Press, 1978), 301.
9. On this point see my "A Pluralistic Account of Space," *International Philosophical Quarterly* 11, 2 (1971):207–12.

8. John B. Cobb, Jr.

WHITEHEAD'S LATER VIEW ON SPACE-TIME:

A RESPONSE TO PATRICK HURLEY

We are indebted to Patrick Hurley for a careful and illuminating study
of the development of Whitehead's thinking about time. There are a
few points in his account of Whitehead's theory in *Process and Reality,*
however, that seem to me in error.

1. The fact that an occasion may share distinct durations with
two occasions that are not themselves in the same duration
is not, for Whitehead, a problem to be solved but a fact
established by relativity physics.

2. The extensive continuum is not an eternal object. Whitehead
distinguishes several types of potentials. *Pure* potentials are
eternal objects. These must be distinguished from impure
potentials, which are propositions, and real potentials, such as
the extensive continuum. The extensive continuum *is,* in
Whitehead's view, extensive. If it were, as Hurley supposes,
an eternal object, then, as Hurley notes, it could not be extensive.

3. Although Whitehead may be more accepting of diagrams
portraying time than was Bergson, he did not affirm the "spatialization of time." The quote Hurley offers to the contrary
states that the spatialization of the world is not an error. That
is, the world is *really* spatial. But the world is also temporal.
Time is one dimension of the world among others. It is not
a dimension of space.

Three other formulations seem to me misleading.

1. It is true that a nexus composed of causally unrelated
occasions must be nonsocial, since a society is distinguished
by the causal transmission of shared characteristics. Hence a
duration is certainly nonsocial. It is also true that the contemporaries of any occasion constitute, as a whole, a nonsocial
nexus, but this is not because they are all causally unrelated
with one another. On the contrary, there are many patterns
of causal relations among them, and there are many societies

that are included. They are nonsocial as a whole because it is not possible that all of them jointly could constitute a single society.

2. Since there is no such thing as a dimension that is nonextensive, the suggestion that becoming is "like" a fifth dimension is of doubtful help to the imagination.

3. Although it is not metaphysically impossible that when an actual occasion appears in the world as a four-dimensional quantum it will be objectified in the presentational immediacy of some other occasion, Hurley himself points out how misleading this statement could be. Presentational immediacy refers primarily to sense experience, and individual four-dimensional quanta or occasions do not, ordinarily at least, enter into sense experience.

II Bohm, Prigogine, and Process Philosophy

9. David Ray Griffin

BOHM AND WHITEHEAD ON WHOLENESS, FREEDOM,
CAUSALITY, AND TIME

David Bohm's passion is to overcome fragmentation. As a reflective person, he is acutely aware of the problems, intellectual and social, that have resulted from the modern vision, which sees all things as externally related to all other things. This vision has led to the assumption that the truth about the world could best be learned by assigning its various "parts" to separate "disciplines," and the relative success of this procedure has reinforced the conviction that the intellectual divisions correspond to real divisions within reality, e.g., between living and nonliving, between mind and matter, between humanity and nature, between deity and the world. This modern vision has reinforced the egotistical and tribal tendencies of us humans to think that the welfare of the individual person or at least group (social, cultural, religious, and/or economic) can be promoted by ignoring (or even defeating!) the welfare of all the others.

As a physicist, Bohm is aware that his own discipline was the major contributor to the rise to dominance of the mechanistic tradition in the modern world, that is, the Western world since the seventeenth century. Furthermore, he knows that physics is still regarded as the paradigmatic "science," implying that overcoming the mechanistic vision could be achieved most effectively if developments in physics itself showed its inadequacy. This is exactly what he believes has happened in the twentieth century, and he has devoted himself to trying to drive home this fact and to working out the new, nonfragmenting vision he believes best makes sense of all the facts and provides a vision of wholeness adequate to our intellectual, religious, and ethical needs.

But these pragmatic considerations do not provide Bohm's only motivation. He is also deeply committed to discovering the *truth* about reality, to the degree that this is possible in our time. This leads him to reject the finality of the present quantum physics, whose equations merely describe the probability of what an observer with a certain instrument would observe, since this means that "modern physics can't even talk about the actual world!" (RV 45; see list of abbreviations at the end of this chapter). Not only does Bohm find this nonrealism unsatisfying, he sees that it keeps the mechanistic vision, which became so deeply ingrained between the seventeenth and the nineteenth centuries, from being effectively challenged. So finally Bohm's two passions coincide: he believes that a *realistic physics,* which will once again intend to express the truth about the world (however partial this truth may be), will point to a *vision of wholeness* in which all things are seen as internally related to all other things.

Because Bohm radically rejects the twofold tendency to see reality as composed of externally related things and to divorce "physics" from "psychology," "philosophy," and "theology"—the twofold tendency that is the essence of the modern vision—Bohm's vision is radically post-modern. (Bohm is *post-* rather than *pre*-modern because he wants to preserve the truths and positive values—and he does not minimize these—that have been attained by *modern* science.) In this paper I shall summarize Bohm's proposal for a post-modern vision of reality (section 9.1) and then discuss some problems involved in his proposal as developed thus far, problems relating to freedom, causality, and time (section 9.2). Finally, I shall suggest how various distinctions within Whitehead's formulation of a vision of "prehensive wholeness" can avoid these problems while retaining the central intuitions in Bohm's vision (section 9.3).

9.1. Underlying Wholeness: Internal Relatedness *via* the Whole

There seem to be two basic ways of explaining how things are *internally* related to each other—so that knowing the truth about one thing would ultimately involve knowing the truth about all things and promoting the good of one thing would involve promoting the good of all things—that would reject the mechanistic view, which sees things as having merely external relations to each other. One way would be *prehensive wholeness,* seeing each individual as a microcosm, somehow grasping all other things into its own reality. Another way would be *underlying wholeness,* seeing all individuals as internally related to all others not directly, but by virtue of the fact that they all arise out of a common ground, which is thereby immanent in each of them, making each of them indirectly immanent in each other. Much of Bohm's language suggests that the second way, that of underlying wholeness, is his. For example, he says that "everything, including mind and matter, actively enfolds the whole (and through this everything else)" (S 39; cf. S 20, 32, 34; RS 333, 337).

Bohm rejects the notion of "interaction," whether between "mind" and "body" or between two "particles." The term *interaction* suggests— and one must admit that Bohm is correct here—that the two things are first what they are, independently of each other, and then enter into relations with each other. The relations would then be external to their respective essences; the relation would not be constitutive or internal to either of them (W 126f., 134, 137). Of course, Bohm knows that there are organismic views, such as Whitehead's, in which inter-action is not thought of in this way, but as involving mutual in-fluence (in-flowing), which is internal to each party. But he also seems to reject the language of interaction because in the vision of underlying wholeness one finite (explicate) thing does not *directly* affect another one at all. Rather, all influence is mediated *via* the implicate order, in the whole. Event *A* arises out of the whole ("projection") and thereby affects the whole (however slightly). Then new events arise out of the whole that appear to have been causally affected by event *A*. But they were not directly affected. Event *A* directly affected only the whole, and the later events each arose out of the whole, so they were only indirectly affected by event *A*. Hence Bohm speaks of the *appearance* of causation and of things behaving "as if" there were a force between them (RV 36; W 184). There is really no "horizontal" causation from surface event to surface event. All causation is "vertical," from the bottom up (projection) and then from the top down (reinjection).

This model can provide the basis for a solution to the mind-body problem, as well as for the wider problem of the "interaction" of mind and matter in general. We do not have to conceive of mind as having a direct influence on matter, or vice versa, but can see that the correlations are due to resonances in the implicate order. Likewise, the nonlocal correlations implied by quantum theory need not be explained in terms of literal "action at a distance" or of supraluminal signals, but can be understood as involving events that arise as explications of resonances in the implicate order (W 129, 186), where the separative space of the explicate order does not exist, except in implicate form: all places are enfolded in the whole. The phenomena of parapsychology that seem to suggest action at a distance (e.g., telekinesis, telepathy) would presumably also be explained by Bohm in this fashion. Even precognitive phenomena, which seem to imply the influence of the future on the present, might be so explained, since Bohm sometimes suggests that the implicate order is timeless, in the sense of enfolding all times (W 155, 167; RV 36). So-called precognition would really involve only the resonance of an event that is explicate *now* with an event that is *later*—from the viewpoint of the explicate order, which orders events sequentially—to become explicated.

The solution to the mind-body problem mentioned in the previous paragraph would imply that what we call "mind" or "experience" or "consciousness" is as fully an example of the explicate order as what we call "matter." Development of this line of thought would make Bohm's position somewhat similar to Spinoza's. Spinoza thought of there being one infinite substance that has an infinity of attributes, with thought and extension being the only two known to us. By denying that there *are* two kinds of substances, and in fact that there is a multiplicity of distinct "substances," Spinoza avoided the Cartesian problem of how two totally different kinds of substances, thinking substances and extended substances, interacted. "Mind" and "matter" are simply two *attributes* of the whole, and attributes are not the kinds of things that have to figure out how to interact.

However, there is another tendency in Bohm's thought, the tendency to say that "mind" or "consciousness" is more illustrative of the implicate order than is "matter" (W 197; S 31, 32). This is a tendency to which I shall return. For now I need to explore the monistic question suggested by the Spinozistic parallel. Regardless of how Spinoza should be interpreted, is Bohm monistic in the radical sense of attributing all agency to the whole, with the result that the apparent multiplicity of individuals have no agency of their own *vis-à-vis* the whole so that a complete determinism (albeit a nonmechanistic one) is the ultimate truth? Some of his language does suggest this, as when he portrays the universe, in Hegelian language, as observing and describing itself through human beings (RS 339), when he says that each event in the explicate order is "simply a projection" of the whole (RV 43), and when he speaks of an "overall necessity" (W 181; cf. W 195, 204f., 209, 213) and suggests that if we could "actually determine all the sub-quantum variables" we would be able "to predict the future in full detail" (W 106). However, he has clearly rejected this interpretation of his meaning. He sees the indeterminism of quantum physics as pointing to inde-terminism as a property of matter (W 85, 105). And he affirms that the universe is "a self-acting whole" that is "in some sense distinct from (i.e., autonomous and independent of) the activity of the entities of the explicate order" (RS 336, 333). This implies that the *entities of the explicate order have some autonomy of their own,* making them distinct from the activity of the whole. Bohm affirms this explicitly, saying that "each of the sub-wholes has its appropriate kind and degree of freedom" and that, because of the law of freedom, the harmony of each event with the whole and hence with all others cannot be perfect (RS 337). However, he does want to insist that the holomovement, the activity of the whole, is *primary* and that the individual events have a "vanishingly small degree of substance or independent actuality" in relation to the totality (S 93; RS 334, 339).

One further point to mention in this brief overview of Bohm's view of wholeness is that he means it to provide a way to explain how novel forms can appear in the explicate world. If events simply arose from the past explicate world, the rise of genuinely new forms would be unthinkable. Also, allowing for novel forms to be inserted now and then by an agency beyond the multitude of finite beings would seem to involve an *ad hoc,* exceptional type of influence. But Bohm's view is that events are constantly being created by the whole and then dissolving back into it. This allows a *natural* way for a creative content to enter the world at any point (RV 47). Apparently enduring things, such as electrons and minds, are really "world tubes" composed of series of events, each event replicating its predecessor more or less exactly (more exactly in the electron, less so in a human mind). The other presupposition necessary for explaining the emergence of novelty is that the whole has a *purpose* to bring about new subwholes. This "deep intent" of nature (RV 39, 40) can explain why the evolutionary process has brought forth a richness of forms far beyond anything survival as the only goal would dictate (RV 40).

9.2. Problems in the Vision of Underlying Wholeness

In this section I shall discuss several problems that arise if Bohm's position is to be understood as that of underlying wholeness alone, without the direct horizontal causation involved in prehensive wholeness. These implications would be "problems" because they would be inconsistent with some of Bohm's deepest concerns and/or with some of our deepest convictions; at least they could be reconciled with them only by *ad hoc* measures.

(1) One problem posed by Bohm's idea of underlying wholeness is that human *freedom,* or the power of self-determination, is minimized. In the first place, Bohm has not given primary attention to the issue of freedom. When he characterizes the mechanistic vision, the focus is always on the externality of relations rather than, as for many people, the determinism implied by mechanism. He even says that the fact that the laws of quantum physics are statistical instead of deterministic has little or no relevance to the issue of mechanism (S 10f., cf. W 173, 178). And, as indicated above, when describing his own position he has not always been careful to avoid statements that could be interpreted to mean that all events, human and nonhuman, are *totally* determined. Second, after clarifying that he sees all events as having some degree of agency for self-determination *vis-à-vis* the whole, Bohm stresses that

the agency of the whole is primary, whereas that of the subwholes is "vanishingly small." Third, focusing on the dialectic between the implicate and the explicate, combined with (sometimes) seeing "mind" and "matter" as equally "explicate," has a leveling effect; it suggests that the human mind has the same "vanishingly small" degree of power for self-determination as an electron: there is no suggestion of a *hierarchy* in nature, with increasing degrees of self-determining power.

This minimization of our agency undermines the conviction running throughout Bohm's writing that the mechanistic, fragmenting vision has been a tragic distortion of the truth, one which we need to and can overcome. If we have only a vanishingly small degree of power *vis-à-vis* the whole, how can we believe that we have *deviated seriously* from its "deep intent"? Here we have a version of the problem of evil: if the creatures have only very little power in relation to the creator, so that they are virtually *mere* creatures (not self-creating ones to any significant extent), how can they significantly "sin," i.e., miss the mark? Also, if our power for self-determination is so minimal, how can we believe that our efforts to develop better insight, and to share this insight with others, can have any effect (even aside from the issue of whether we can affect others directly, or only *via* the whole)? To stress that our power is vanishingly small is implicitly to say: whether or not a new vision comes to dominate is primarily up to the whole, hardly at all up to us. This belies the passion involved in Bohm's own efforts to help change the dominant vision. (It may be true that true insight comes not from effort in the usual sense, but through being receptive to inspiration. But, even to the degree that this is true, it takes considerable effort to get ourselves into a truly receptive attitude!)

(2) A second set of problems arises from Bohm's apparent denial that events exert any *direct causation* upon other such events. One problem here is simply that this denial would run *counter to one of our deepest convictions,* which is that we do directly interact with other things from which we are partly distinct—that other things do affect us directly, and that we directly affect other things. My body affects me, and I my body; through my body I am affected by the surrounding world, and I affect it. Bohm's formulation—according to which each enduring thing is a series of events, each of which arises from the whole and then dissolves back into it, thereby modifying *it* slightly— seems to deny this conviction. Each event affects other events *only* by affecting the whole, out of which the later events arise. It is similar therefore to Malebranche's view: I cannot kick you directly, but only (as it were) by kicking God who in turn kicks you!

The denial of direct effects would also make the *stability* of the world mysterious, reconcilable with the theory only by an *ad hoc* solution. Why do certain forms of order, e.g., electrons and molecules, keep repeating themselves for eons? Bohm admits that his view entails that "in principle, every new moment could be entirely unrelated to the previous one—it could be totally creative" (RV 36, cf. W 205). But experience shows us, as he points out, that "there is usually a great deal of recurrence and stability leading to the possibility of relatively independent sub-totalities" (W 205). This idea of "relatively independent sub-totalities" is stressed repeatedly by Bohm; it connects his views with the world as experienced by us. And he does seek to explain how this occurs in a way that is consistent with his basic principles, by suggesting that a series of repetitions of a form will create a "disposition" of the implicate order to produce that form (RV 36). A form is projected into the explicate order, then introjected back into the implicate order, then back into the explicate, and so on. Each introjection influences the whole, creating a tendency for it to explicate itself in terms of that form (RV 36). This is how Bohm explains "the appearance of the 'causation' of the present by the past" (RV 36) and the "interesting point" that "each moment resembles its predecessors" (RV 42).

However, one thing this theory does *not* explain is why the same (or similar) forms are almost always repeated in roughly the same place, *vis-à-vis* the other forms that are being repeated. The forms embodied in the "aggregate of events" I call the typewriter before me tend to be repeated second after second, minute after minute, hour after hour, in the "same place," i.e., with the same spatial relations to the other forms that are being constantly repeated, *viz.*, those I call the house, the desk, the lamp, and my body. Bohm's account seems at most to account for the disposition of the whole to repeat the same forms; it *does not account for the disposition to repeat them in the same or at least a contiguous place.* Bohm's theory has the virtue of explaining how the phenomena normally called "teleportation" or "materialization and dematerialization" can occur. But on the basis of his theory we should expect these phenomena to be much more common than experience teaches us they are. If events do not *directly* affect their successors in a world tube, but only by first influencing the whole, in which all times, places, and forms are merged, we have no reason to expect the introjected form to be reprojected next to approximately the same forms it was near in the previous moment.

Another problem is closely related to this one: if one thing does not really affect another one directly, but only indirectly *via* the whole, why is causation between *contiguous* events so overwhelmingly important in our world? Bohm's theory explains how nonlocal correlations in physics (and parapsychology) can occur. But Bohm himself says that the evidence to date in quantum physics suggests that nonlocal effects arise only under very special conditions. Furthermore, even if physics does come to show that *all* "particles" manifest nonlocal correlations with others, and if parapsychology convinces us that there are influences (or at least correlations) between noncontiguous events far beyond the relatively few instances of *consciously* detected extrasensory perception and *obvious* psychokinesis, this will not change the fact that causal relations between contiguous events are overwhelmingly important. Bohm's theory, by saying that every event is connected with every other event in the implicate order, explains how nonlocal correlations in the explicate order are possible. But *if it denies that events have direct influences upon other events, it does not explain why local correlations are so important.*

Finally, there is an element of *arbitrariness* in the affirmation that events exert direct causal influence upon the whole (RV 36) but not directly upon other subwholes. Insofar as Bohm distinguishes between the whole and the multitude of subwholes, allowing each some autonomy *vis-à-vis* the other, it would seem more consistent to allow that each event would have a direct influence upon subsequent subwholes, as well as an *indirect* influence upon them *via* its influence upon the whole. This vision would combine underlying and prehensive wholeness.

(3) Another set of problems could be created by Bohm's suggestion that all *times* (as well as places) interpenetrate in the implicate order (RV 36; W 155, 167), which implies that the implicate order would be beyond time (RV 37, 43) in the sense that the *distinction between past, present, and future would not be real* in that order.

In the first place, there is the problem of the compatibility of the belief in genuine freedom and creativity, which Bohm wants to affirm, with the belief that in the "really real" order the totality of what *we* regard as future is already settled. If those events which are still future for us are, in some more real realm, as fully settled as those which we regard as past, that is, *if the present implies the details of the future as fully as it implies the details of the past, then each present event really has no power of self-determination.* Accordingly, insofar as people are logical, belief that there is an implicate order in which all times interpenetrate will undermine Bohm's call to exercise our creativity to change the way we and others think.

In the second place, Bohm thinks of the implicate order as having awareness and purpose (RV 37, 39). To think of the implicate order as "the totality beyond time" (RV 43) raises all the problems that have been endlessly debated as to *whether it is even meaningful to speak of a "nontemporal awareness" and a "nontemporal purposiveness."* I say that these have been endlessly debated; but there does seem to be a growing consensus, shared by atheists (e.g., Sartre) and theists (e.g., Hartshorne) alike, that these notions are not meaningful because they are self-contradictory. Bohm is aware of the problems. He says: "Whatever knowledge this implicate order would have would be beyond time. Therefore, I don't know if you would even think of it as knowledge" (RV 37). And after saying that the universe seems to be experimenting with forms, he says: "it shows itself to us *as if* it were experimenting. That is, when looked at from the limited aspect of time, the structure *looks like* an experiment" (RV 37; emphasis added). If those few events which seem to imply that the future is already as settled as the past, so-called precognitive events, could be explained without denying the ultimate validity of the distinction between past, present, and future (and I argue elsewhere that they can), would it not be better to limit the nontemporality of the whole to an abstract element within it and to retain the *experienced* asymmetry between past and future as an ingredient in our theories about the ultimate nature of things, since we can thereby more clearly retain freedom and meaningfulness?

9.3. Whitehead's Vision of Prehensive Wholeness in Relation to Bohm's Concerns

Some of the problems in Bohm's formulation of his vision thus far are matters of self-consistency; others are tensions between his formulation and our deepest convictions. All these are related, I suggest, to a more general problem, a tendency to use the implicate-explicate distinction too indiscriminately. Many of the analogies Bohm has lifted up between apparently dissimilar features of the world and human experience are indeed illuminating. But in some cases the general formula, "the explicate unfolds from the implicate and is then enfolded back into it," leads to tensions with our deepest intuitions or with other applications of this formula. These tensions arise because some fundamentally different types of relations are subsumed under the general implicate-explicate formula.

In this section I suggest that some distinctions developed by Alfred North Whitehead can preserve the affirmations about which Bohm is most concerned while avoiding the problems in his formulation of these affirmations as developed thus far.

Whitehead in many respects traveled a path similar to Bohm's. He also was a mathematical physicist (though primarily a mathematician) passionately interested in the relation between the world as described by physics, on the one hand, and the phenomena of life, on the other, and the relation of both of these to the world as known through moral, aesthetic, and religious experience. He was also deeply disturbed by the deleterious effects the vision of reality formulated in relation to the natural sciences has had upon the modern world (see his *Science and the Modern World*). He likewise began with a vision of wholeness reminiscent of Spinoza's but then went beyond it. It is primarily the distinctions he developed in moving beyond the Spinozistic monistic and deterministic vision that are helpful in resolving the tensions within Bohm's developing position.

9.3.1. God and Creativity

Much philosophical and religious thought, both Eastern and Western, has understood undifferentiated being as the ultimate reality. It has variously been named Being, Being-itself, Prime Matter, *Urgrund,* the Godhead, Brahman, and Emptiness. Whitehead calls it Creativity. It is formless; it is being without attributes (Nirguna Brahman). Whitehead says "it is without a character of its own" (PR 31).

There have been two major ways in which this metaphysical ultimate has been thought to be related to a determinate, perhaps personal, deity (Saguna Brahman). On the one hand, a determinate deity has been regarded as the first *emanation* from the indeterminate ultimate reality. *How* that which is totally devoid of all form, all determinateness, could give rise to something *with* form has always been a problem, but the affirmation has been widely made.

On the other hand, some traditions have simply identified God and Being-itself. This creates inevitable tensions. Sometimes, as in Tillich, the affirmations made about the personal God of religious devotion have to be interpreted so as not to contradict the philosophical vision of an ultimate reality said to transcend all determinate characteristics and hence to be beyond attributions of, e.g., love, knowledge, purpose, and agency. The pious are allowed to continue applying such terms to Being *qua* God, but the philosophical theologian knows that the attributions cannot be applied literally, or even analogically, but only symbolically, metaphorically. This view also means that "God" can exert no influence: there is no concrete "whole" in the sense of an all-inclusive embodiment of Being. The only embodiments are the multiple finite instances of Being. Being-itself is not *a* being that can influence or be influenced by the various beings; it is simply the being of the many beings.

At other times, as in Thomas, the indeterminateness of Being-itself is compromised by being equated with the determinate God. This equation leads to total determinism, since *all* being and hence power and activity belong to (are identical with) God.

One of Whitehead's major innovations was to diverge from these two dominant ways of relating God and Being. Whitehead *distinguishes* between God and Creativity and yet makes them *equally* primordial. God is not simply Creativity; God has determinate characteristics: God knows the world, envisages primordial potentials with appetition and purpose, influences the world, and is in turn influenced by the world. God loves the world actively, seeking to influence it toward its good, and receptively, responding sympathetically to its events. But God is not a derivative emanation from Creativity; God is the primordial embodiment of Creativity. (Creativity is that which is instantiated by all actualities; it is not an actuality that could exist by itself, unembodied.) Whitehead refers to God as the "eternal primordial character" of Creativity (PR 225, cf. 344).

I suggest that Bohm has thus far wavered between these three visions. Sometimes he speaks in a Vedantist-Neoplatonic way, as if the ultimate reality, the ultimate implicate order, were totally formless. For example, he says in an interview: "We must have some form—we can't live entirely in the implicate order" (RV 36). In this mood, he speaks of all "measure" as created by human insight, *denying* that "it exists prior to man and independently of him" (W 23). Reality as such would be formless, Brahman without attributes. All form and measure would be *maya,* illusion. In line with this vision, Bohm can legitimately say that we have freedom, for each of the explicate parts, each of the events of the world, would be an embodiment of the whole, which is a holomovement, dynamic activity. But if he were to carry out this vision consistently, he would not be able to talk of the influence of the whole on the parts (except as the "whole" in the sense of the totality of the parts, and this is the mechanistic vision he wants to avoid), nor the influence of the parts back upon the whole (i.e., of the enfoldment back into the implicate order as somehow altering it). It is not for nothing that consistent visions of this sort stressed that the ultimate reality was impassible.

More characteristically, Bohm seems to equate God and Being somewhat in the Thomistic fashion (at least as I am interpreting Thomas) and to see this somewhat determinate reality as the ultimate implicate order. Accordingly, Bohm speaks of the ultimate implicate order as having intelligence and compassion. But this vision, if carried out consistently, would lead to determinism, for if all energy, movement, or activity as such is equated with a concrete being (and only a concrete, determinate being can have attributes such as intelligence and compassion), then the creatures have none of their own. In this vision, *all causation is vertical:* there is a hierarchy of levels of order. Each level (except the highest and the lowest) is implicate in relation to the level above it and explicate in relation to the level deeper than it. (In line with speaking of "underlying wholeness," I am referring to the more implicate orders as *deeper,* even though Bohm often speaks in Neoplatonic fashion of descent from the *highest* implicate order to the more explicate orders.) The "implicate-explicate" language here suggests determinism: each level is a mere explication or unfolding of what was already there, implicit or enfolded, in a deeper level. Indeed, Bohm sometimes says that there is an infinite hierarchy of implicate orders, suggesting that this somehow avoids the conclusion of total determinism. But this is problematic. First, it is hard to see what it might mean. Second, it is hardly consistent with speaking of the *ultimate* implicate order as characterized by love and intelligence. Third, if the level of conscious human experience is totally a product of some deeper level, it does not mitigate the implied determinism to say that the series of increasingly deeper levels of causal orders never reaches bedrock. But the fact that Bohm thinks there is a problem requiring a solution shows that he often does not think of each level of reality as having its own activity, creativity, or freedom, by which events can partially determine themselves *vis-à-vis* other levels and the whole.

Bohm's statement that these events do have some such power, but that it is "vanishingly small," can be regarded as a compromise between the first two visions. But his intuitions that these explicate events somehow affect the whole, even if only slightly, can fit with neither vision. If the ultimate implicate order is *formless,* we cannot affect its form; if it has *all* the activity, in which case we are merely emanations from or explications of it, then we have no agency by which we can effect a change in it.

But Bohm's intuitions here can be conceptualized in terms of Whitehead's new vision, in which God and Being (i.e., Creativity) are distinguished. Creativity is embodied by all events. Creativity is the threefold capacity of events (1) to be influenced by previous events, (2) to create themselves partially, and (3) to influence subsequent events. Creativity is embodied by every local event ("actual occasion") and by the all-inclusive series of events ("God"). Accordingly, God has autonomous power to influence the world and the capacity to be influenced by it (this latter capacity is called God's "consequent nature"). Likewise, each local event constituting the world arises from the Whole at that moment, meaning God and the totality of previous actual occasions. But each event then influences God and all subsequent local events. Accordingly, the Whole out of which the next moment of the world will arise will be slightly different from what the Whole was a moment before.

In this vision, God is not seen as owning Creativity (or Being-itself) any more than does the world. God has always existed, instantiating Creativity. But so has the world—not *this* world, with its contingent forms of order, but *some* world or other, with a multiplicity of actual occasions embodying Creativity. Creation of our particular world was not initiated by a creation *ex nihilo,* in the sense of a total absence of finite forms of actuality, but was a creation out of chaos, out of a less ordered realm of finitude.

Accordingly, the relation between God and finite events cannot be described in the language of implicate-explicate, for at least two reasons: (1) God and local events each have self-determining power in relation to each other, so neither is merely the unfolding of what was contained implicitly in the other; and (2) local events are directly influenced by previous local events, not just by God: the Whole out of which they arise is God-and-the-world.

Although there *is* hierarchy in this Whiteheadian vision, there are distinct realities that are *not* related hierarchically. Rather than a Neo-platonic-type descent from Creativity, to Forms, to God, to Creatures, all of these realities are *equally* metaphysical, equally primordial (with the qualification that no particular creatures are necessary, only Crea-turehood as such; i.e., there must be *some* creatures). God is as pri-mordial as Creativity, each implying the other. God, as a determinate being, can act: God's primordial activity is the appetitive envisagement of the Eternal Forms ("eternal objects"), the Primordial Potentials, which imply God and Creativity as much as being implied by them. And God, Creativity, and the Forms all imply, and are implied by, a realm of Creatures who will in-form their Creativity with a selection from the Eternal Forms.

This set of mutually implied (rather than hierarchically arranged) realities protects our intuitions about freedom, causation, and time. The distinction between God and Creativity, which allows the creatures to be equal stockholders in Creativity, protects the freedom of the creatures. It also protects the concreteness and transcendence of God and, hence, causal influence between God and world. The idea that God is not derivative from Creativity protects the ultimacy of determinateness, including the temporal distinction between past, present, and future. The idea that the Creativity embodied in the creatures is not derivative from God also protects the ultimacy of temporal distinctions. The idea that God also embodies Creativity gives further support to the ultimacy of temporal distinctions, since, besides God's primordial, nontemporal aspect, there is God's concrete actuality, which is temporal, i.e., which distinguishes between events that have occurred and possible events, which have not. Also, this fact that God as concrete is temporal allows us to speak of an all-inclusive awareness and purpose without contradiction and without undermining the reality of time and freedom. Our future could hardly be indeterminate, to be rendered determinate only by our exercise of creativity, if from a higher point of view the events that seemed future to us at a certain "now" were already (or eternally) determinate.

9.3.2. Events in Themselves and for Others

I have distinguished between three phases of Creativity: an event's (1) reception of influences from its environment, its (2) self-determining activity, and its (3) influence upon subsequent events. In this section I shall collapse the first two moments into one and refer to this moment as the event as it is *in itself;* the third moment will be identical with the event as it is *for others.*

The event in itself is a subject. It does not enfold the influences from the environment the way a cabinet receives canned goods, but the way a moment of experience receives influences from its body and the greater world. It does it *with feeling.* In fact, Whitehead refers to each local event, each "actual occasion," as an "occasion of experience." Every true individual (as distinct from aggregates of individuals, such as sticks and stones) has (or is) a unity of experience in which a vast myriad of influences are synthesized. This reception of influences, and self-determining synthesis of them into a unified experience, is what an event is in itself. This internal, self-determining process is called "concrescence," which means "growing together." This notion corresponds with Bohm's attribution of an inner formative activity to events in their phase of enfolding (W 12, 13, 79).

But as soon as this unity is reached, the event becomes an object for others. The subject becomes a "superject." The event as a becoming subject "perishes"; the event as a causal power upon others comes into being. The data it had enfolded are now unfolded or "superjected" into the universe. The event reveals publically what it had been doing privately. What was a subject in itself becomes something for others, and *in this sense* it is an object. Whether it also becomes an object of sense perception, and/or an object of consciousness, depends upon whether there are subjects around capable of making it into an object in this more sophisticated sense. But it, willy-nilly, becomes an object in the sense of a causal influence upon subsequent subjects, which unify it (along with the rest of the environment) into their internal reality and which then in turn perish as subjects and become objects or superjects for subsequent subjects, and so on.

This provides a way of distinguishing between mind and matter without relying on an ontological dualism making their mutual influence unintelligible. Rather than being two different types of actualities, "mind" refers to what an actual entity is in itself, whereas "matter" refers to what it is for others. Our self-consciousness at a moment is our direct knowledge of what an event is in itself: we know what a "thing in itself" is by *being* one—and by being one that is sophisticated enough to be aware of itself. (Lower-grade subjects would have awareness, but not self-awareness.) We do not have the same kind of direct knowledge of the subjectivity of other individuals. But since we know that *we* have subjectivity, even though it does not appear to others, we can assume, by analogy, that other individuals had their own subjectivity prior to their becoming objects for us. Bohm suggests this nondualist position, saying that what we call "matter" has something analogous to mentality, creativity, and imagination (RV 39, 47).

I am using the word *individual* deliberately. There are also *aggregates* of individuals, such as sticks, stones, and tables. These answer to the ordinary notion of matter even more than does the objective existence of individuals such as electrons, atoms, molecules, and cells. These aggregates show no signs of the spontaneity we associate with subjectivity, since the uncoordinated spontaneities of the millions or billions of members of the aggregate (e.g., the molecules in a rock) cancel each other out with the result that no unified movement is attained.

There are some groupings of individuals that are not mere aggregates, however. These are "compound individuals," in which a higher-level series of subject-then-object events arises and has a dominating influence over the society as a whole. Animals, including ourselves, are the obvious examples. But atoms, molecules, and living cells can also be thus regarded. The world of finite things can then be classified into these four basic types of things: (a) actual occasions; (b) enduring objects, which are serially-ordered societies of occasions (Bohm's "world tubes"); (c) nonindividualized societies of enduring objects (inorganic things, plants); and (d) compound individuals. A high-level compound individual harbors, as its dominant member, a series of higher-level occasions of experience, a "soul."

It seems to me that this position is already implicit in Bohm's thought, insofar as he speaks of each event as enfolding the whole and then unfolding itself back into the whole and of a so-called particle as in reality being a world tube or a trajectory of such events of enfolding and unfolding. His complaint against orthodox physics would be that it has thus far assumed that the events in their unfoldment, or explicate state, constituted the full reality of the events, thus ignoring their prior, implicate state. His suggestion that there are "hidden variables" to account for the behavior of observed events would mean that the explanation may lie in what the event is in itself, in its subjective moment, which has at least an iota of self-determining power. Bohm comes close to this view when he suggests that mind or consciousness is more illustrative of the implicate order than is matter (above, section 9.1). I am urging him to say that self-conscious experience is our one opening into what an event in its state of enfolding is and then to generalize some degree of experience to all events.

The Whiteheadian position would have the following advantages to Bohm:

(1) It would show how mind and body can be directly related, without having to route this apparently direct relation through some underground reality. Bohm is right to say that mind and matter are related through some more fundamental reality, but this more fundamental reality need not be thought to exist beyond the concrete events of the world. The concrete events are themselves this more fundamental reality, each being first a subject which enfolds previous subjects-become-objects into itself and then in turn unfolds itself as matter for subsequent events. What we call our own "mind" or "soul" is simply a very high-level series of subject-object events dominating a body made of societies composed of lower-grade events with this dual nature. Since there is no ontological dualism, mutual influence is no problem.

(2) This position would show the fundamental reality, and irreversibility, of time. Time results from the causal relations between events. The irreversibility of time is due to the relationship of enfolding and being enfolded. If event B is later than A because B included A but A does not include B, it would be nonsense to suppose that time could then go backwards, so that event A would be later than B. For this would mean that B would both include and not include A, which is a self-contradiction.

Time is not an actuality which could exist apart from events. But since Creativity is the ultimate reality, so that there always has been and always will be ongoing relationships of including and then being included, time in the sense of a distinction between past (determinate, included), present (including, becoming determinate), and future (indeterminate, not included [at least in the same sense as the past is included]) is a necessary feature of reality.

(3) This doctrine allows real causation (vs. positivism) but without a mechanistic view of causation as total determination and mere external relation. The direct causation of one event upon another can be affirmed, and yet Bohm's view that there is no causation between two explicate objects is supported. The causation of one event (as object or superject) does not directly influence a subsequent event's superjectivity; rather, it passes through the affected event's subjectivity and hence through an implicate ordering process, which involves some element of self-determination and which is hidden to the outside observer. It is only when the affected explicate event is our own behavior that we are privy to what goes on *in between* the two explicate events.

Of course, we are only *partly* privy to this process, since much of it transpires below the threshold of consciousness. This is another way in which Bohm uses the implicate-explicate distinction: conscious experience can be considered an explication of unconscious experience (S 67). One moment of conscious experience, as explicate experience, does not directly affect a subsequent conscious experience, but only indirectly, by passing through the unconscious depths of the next moment of experience. It should be noted, however, that this is not an example of causation by one event upon another, but of different phases within a teleological, self-determining process. But even here the explicate (i.e., conscious) aspect is not *merely* an explication of what is already determined in the implicate depths: the conscious aspect of experience plays a role in the self-determination of a moment of experience. Consciousness is not merely an epiphenomenal by-product of unconscious forces. So, even though the implicate-explicate formula works here better than for many distinctions, it is not fully appropriate.

However, terminology aside, Bohm's point is important, and it is supported by Whitehead. Consciousness, according to Whitehead, tends to light up only the later phase of an occasion of experience, not the early phase, where the enfolding of the environment occurs. Hence consciousness tends to lose sight of the connectedness of experience with its world—the fact that it arises from and even includes the whole past world (and God) in itself. Accordingly, the soul, insofar as it identifies itself with its conscious experience, comes to see itself as an independent substance, only externally connected to the surrounding world. Solipsism can even be seriously entertained. Bohm is right: our conscious experience can seem to be even more disconnected from its environment than the matter we perceive (and construct—see below) through our sensory experience (S 94f.). Insofar as Bohm is using the implicate-explicate distinction to stress that this apparently disconnected consciousness is part of a far vaster experiential process in which the whole world is enfolded, the distinction is justified.

(4) Whitehead's way of speaking of an enduring object as a serially-ordered society of events provides a basis for conceptualizing another of Bohm's concerns, which is to affirm that nature has a "deep intent" to realize new forms (above, section 9.1) and that the world is somehow able to respond to this. Bohm says, for example: "You might suppose, say, that somehow nature realizes that it's being presented with various things that now have to be brought together. Nature realizes this greater whole at a deeper level, which is analogous to imagination" (RV 47).

Whitehead's explanation is as follows. The primordial nature of God, which is the Divine Eros of the universe, is God as envisaging the primordial potencies with appetition that they be realized in the world (PR 32–34). Worldly events, for which a given potentiality is relevant, come to feel this potentiality for its future successors conformally, i.e., with appetition. Many successive members of a given enduring society (world tube) could continue to feel this possibility for the society's existence, but as long as it was only felt with appetition, or "mentally," nothing would be changed in the outer appearance of the enduring object. The successive occasions of experience would in themselves be different, in that the new possibility would be fermenting in them; but their outer demeanor would remain unchanged. The new form would only be implicit in the society. But at some point, as Bohm says, "it unfolds into the external environment" (RV 47). In Whitehead's terms, it unfolds when some member of the society feels the possibility not only mentally, which is a restricted way of feeling it, but physically, or unrestrictedly (PR 291). This occurs when there is a "hybrid physical feeling": an occasion feels physically what was felt by a predecessor only mentally. To feel it physically is to accept it as characterizing one's one shape. At this moment, the new form becomes observable and has effects upon the environment. Accordingly, Whitehead supports Bohm's intuition that novel forms do not suddenly arise in the observable world from nothing, or even directly from other observable events, but from an implicate, hidden dimension of the world.

Incidentally, this doctrine is germane to the problem of evil. God's causal efficacy does not directly produce the form of the observable world, but is twice removed from it. First, God must wait for events in the world to feel the divine appetitions for them conformally, i.e., to develop an appetition for these new forms. Second, even after this appetite is whetted, the novel possibilities will not become manifest in the world until some event makes the leap from entertaining the possibility to actually living in terms of it.

(5) The distinction between hybrid physical prehensions and pure physical prehensions provides a causal basis for accounting for nonlocal correlations (i.e., correlations between noncontiguous occasions) while accounting for the special significance of causation between contiguous events. In a *pure* physical prehension, the form prehended from a previous occasion is one that is energized by the creativity in the physical pole of that occasion. Forms that have been physically realized are *unrestrictedly* realized, and hence are superjected with the full energy of the occasion behind them, and hence with considerable compulsiveness. Forms that are realized only conceptually (mentally, appetitively) are not embodied in the event's physical energy, but only in its mental energy, which may be negligible. The prehension of such a form from an antecedent occasion will mean, at least usually, the reception of data without compulsive, but only persuasive, power.

This distinction should correspond to Bohm's suggestion of two forms of energies: the denser explicate energies and the subtler implicate energies, which "would not ordinarily even be counted as energies" (RV 39, 44). Here again, Bohm's "implicate" would correspond with Whitehead's "conceptual" or "appetitive."

Whitehead suggests that *pure* physical prehension is limited, at least for the most part, to *contiguous* occasions. If so, i.e., if compulsive influence occurs only between contiguous occasions, this explains why contiguous causation is so important in our world, so much so that the modern mind has thought it to be the only kind of causation. But since Whitehead does not consider it the only kind, he has a basis for explaining action at a distance. That is, he suggests that *hybrid* physical prehensions can occur equally between contiguous and noncontiguous occasions. This provides for another kind of influence, different from the "physical energy" of current physics.

Hence, Whitehead explains both local and nonlocal correlations in terms of prehension, and hence in terms of direct causation of one event upon another. With a Whiteheadian basis, one need not resort to noncausal synchronicity, rooted in some timeless dimension in which all things are together, in order to explain parapsychological events or nonlocal correlations in physics. By allowing for direct prehensions of remote events, the speed of light does not put an upper limit on the time in which one remote event can influence another. Accordingly, one need not assume that some connection, other than a direct one between the two events, is needed in order to explain the nonlocal correlations in physics and parapsychology. One thereby avoids the problem as to why, if events do not directly influence other events, the apparent causal connections between contiguous events are different in kind from those between noncontiguous events.

Incidentally, I should add that Whitehead was not dogmatic about limiting pure physical prehensions to contiguous events. "*Provided* that physical science maintains its denial of 'action at a distance,' the safer *guess* is that direct objectification is *practically* negligible except for contiguous occasions; but that this practical negligibility is a characteristic of the present cosmic epoch, *without* any metaphysical generality" (PR 308; emphasis added to highlight the four qualifications given). Accordingly, if physics or paraphysics seems to require it, e.g., for instances of psychokinesis, there would be no metaphysical reason to deny that the more compulsive type of causality could be exerted at a distance.

In any case, assuming for the most part that pure physical prehensions occur only between contiguous events, we can see that each kind of causation has its own advantage. The kind exerted only between contiguous events has much more immediate strength, but the kind that can be exerted between noncontiguous events can develop strength through repetition. Hence, Whitehead's position provides a basis for the kind of point made by Jung and Sheldrake, that the *repetition* of a form countless times in the past creates an "archetype" or "field" that exerts a formative influence upon present events. If a particular form is repeated in events *A–D*, event *E* will receive the same form directly from *A*, from *B* (which includes *A*), from *C* (which includes both *A* and *B*), and from *D* (which includes *A, B,* and *C*) (see PR 56, 226, 284). Hence, even though noncontiguous causation of event *A* upon event *Z* may be trivial compared with its contiguous causation upon event *B*, the noncontiguous causation received by *Z* may be as important as the contiguous causation, due to the cumulative effects of countless repetitions of a similar pattern.

This point, although often overlooked by Whitehead interpreters, is a key to his central description of Creativity, which is that "the many become one, and are increased by one" (PR 21). The "many" out of which any event is created is finally the whole past, not just the immediate past: "the whole world conspires to produce a new creation" (RM 109). This view provides a very strong notion of wholeness, of each event as a microcosm, incorporating in some sense the whole world, and of the world as an organism, in which the whole enters into every part, not a machine, in which causation by contact is king. And of course the whole that is prehended by each part is not only the whole past world but also God, who has incorporated the whole past world into an all-inclusive experience. And yet all this wholeness is affirmed on the basis of the category of prehension alone, which means that the distinction between past, present, and future is never compromised. Wholeness does not require giving up the limitation of efficient causation from past to present. The future and the contemporary worlds are left as indeterminate and hence as exerting no causation upon the present event. This allows our intuition of freedom, in the strong sense of the capacity for self-determination in the moment, to remain unthreatened by our intuition of wholeness.

9.3.3. *The Worlds of Causal Efficacy and of Sensory Experience*

One of Whitehead's chief concerns is to distinguish the actual world, in which real causal efficacy is exerted among events, from the world as it appears to our sensory perception, especially vision. This latter "world" is not the world as it actually is, but an appearance produced by our sensory and conscious experience out of the actual world's causal efficacy upon us. This appearance is not a total falsification of the actual world, but it involves gross simplification and distortion. In particular, it presents us with a world in which things appear to be passive rather than active, to be externally rather than internally related to other things, to have no experience, no aim, no self-value. And of course natural science has largely limited itself to this world of appearance—to the world as known through our senses and instruments designed to amplify them. Accordingly, if the world as it appears to scientific study is taken to be the actual world, we get a picture of the world as made of externally related, passive, aimless, valueless bits of stuff. And such a world can clearly provide no intelligible explanations as to why it behaves as it does. Explanation, as opposed to merely descriptive generalization (which is positivism), requires resorting to something hidden beneath the appearances. In the modern period, the dominant assumption among those seeking explanations has been that the actual world is composed of entities whose reality is exhausted by their appearances, their effects. What they are in themselves is not thought to be *essentially* different from what they are for others. This assumption has produced the materialistic-mechanistic world view.

Whitehead, however, decided that actual entities in themselves were subjects, aiming at and realizing value and being internally related to other actualities in their environments. He does not base this conclusion on pure speculation, but upon experience, but it depends upon not taking sensory perception as the basic type of perception. More basic is "perception in the mode of causal efficacy," which involves a direct perception and internalization of other events. At this level of experience, we perceive the actuality of other things, their activity, and some of the values they have achieved. At this level perception is not of an object as external; rather, it is a "prehension," a grasping of aspects of the object into oneself as material for one's own experience. Hence, Whitehead speaks of a direct interaction among events, but he does so without using the image of external, mechanistic relations that the term *interaction* often conjures up.

From a Whiteheadian point of view, Bohm is absolutely right that there is no causation between "explicate" events, *if* this term is used to refer to *events as perceived by sensory experience,* which is how Bohm often uses it (W ix, 158, 186, 206; S 92). That world is entirely a product, not a producer. (It has "effects" only insofar as we act upon the illusory belief that it is the actual world.) And yet, by distinguishing between it and the actual world, we can affirm our deep intuition that there is a direct causal relationship between events, such as between one moment of experience and the next or between mind and body.

9.3.4. Various Levels of Actuality in Compound Individuals

One of the points I have been making is that, although there are several features of Whitehead's vision that correspond to distinctions Bohm makes between implicate and explicate orders, these features do not form a hierarchy of levels of existence, a great chain of being, ending in a deepest level in which everything in the higher levels was already implicit. However, there *are* some types of hierarchy in Whitehead's thought. These also correspond to notions suggested by some of Bohm's statements.

First, there is hierarchy involved in all compound individuals. The atom is already a hierarchical society, since it is not merely an aggregate of subatomic parts. Rather, inclusive of these parts there is a series of atomic occasions of experience that make the atom into an integrated whole. Molecules can likewise be thought to be unified by molecular occasions of experience. The same can be thought to be true of macromolecules, viruses, etc. The living cell is dominated by living occasions of experience. Finally, the multicelled animal is not just a democracy of cells, but has a dominating member, the series of experiences constituting the soul.

Now, when we make a conscious decision, the causation involves all these levels—besides the previously mentioned fact that conscious experience arises to a great extent from integrations made at a preconscious level. This preconscious experience involved enormously complex integrations of data from various parts of the brain. These parts of the brain are composed of brain cells. The functioning of the cells is partly determined by that of their organelles, etc., and those in turn by their molecules, and those in turn by their atoms, and those in turn by their subatomic constituents. Bohm uses this example as an illustration of the fact that what is implicate in relation to a higher level (e.g., the functioning of brain cells is implicate in relation to the person's conscious decision) is in turn explicate in relation to a lower level (the functioning of the cells is explicate in relation to that of their molecules). This surely points to an important truth.

However, it does not provide an example of a kind of "implicate order" that modern science has overlooked. This attempt to explain the functioning of organisms in terms of their elementary constituents has been at the heart of the reductionistic drive of modern science. So this way of employing the implicate-explicate distinction does not provide a parallel to the kind of implicate order that is in principle hidden to the current methods of modern science.

Furthermore, to use the hierarchy in a compound individual as an example of levels of implication suggests reductionistic determinism. To say that our experience is (merely) an "explication" of what was already implicit in the brain is to reduce the mind to the brain. And if the functioning of the brain cells is likewise said to be an explication of that of their constituent molecules, and so on down, the logical conclusion is that human experience is in principle totally explainable in terms of the functioning of subatomic particles. (And then whether they are thought really to be materialistic particles or world tubes of momentary enfoldings and unfoldings is irrelevant, at least to the issue of human freedom.)

In Whitehead's portrayal of the compound individual, the terminology of implicate and explicate orders would not be appropriate for the relation between any two levels. The key point again is the universality of Creativity. Individuals at every level have their own degree of Creativity, and hence power for self-determination, *vis-à-vis* the influences upon them (from above, below, or across). So individuals at *no* level are mere explications of what was already implied at some other level. In fact, far from being reductionistic, Whitehead's view implies that individuals at the higher levels have more creative power than individuals at lower levels. For example, although it may be that brain and mind (or soul) have about the same degree of influence upon each other, the soul is at each moment *one* occasion of experience whereas the brain is composed of billions. This suggests that the soul-experience has billions of times more creative power (to determine itself and then others) than an individual cell-experience. Accordingly, from this perspective, it would be very misleading to suggest epiphenomenalism by seeing each higher level as an explication of a lower one. (And it would also be erroneous to speak, with Christian Science and other forms of hypophenomenalism, of the lower levels as mere explications of the higher, as if cancer always resulted from a screwed-up psyche and never from a polluted environment.)

9.3.5. *Electromagnetic, Geometric, and Extensive Societies*

The point of the above discussion is that the hierarchy involved in a compound individual, such as a human being, is a *hierarchy of actualities,* and as such it cannot be simply a hierarchy of implication and explication, since *all* actualities embody self-determining Creativity. However, there is another kind of hierarchy of societies in Whitehead. Whitehead suggests that all compound individuals (discussed above) are specialized societies (developed to foster more intense experiences) within more general forms of social order. These latter societies are related in a somewhat Chinese-box style, with the less general orders being totally included in more general ones.

The first level of generality, he suggests, is the electromagnetic society, composed of electromagnetic occasions. The order of this society has physical relationships determining the importance of one family of straight lines, one definition of congruence, and systematic law—which is statistical (PR 98). This society is our present "cosmic epoch." It is set within a far wider society, the geometric society, which has those relationships which make possible definable and hence measurable straight lines. But there can be competing definitions of straight lines and hence alternative systems of metrical geometry, so this geometrical society could be patient of cosmic epochs quite different from our own electromagnetic one. And this geometric society is set in turn within a far vaster society of pure extension, the properties of which express "the mere fact of 'extensive connection,' of 'whole and part,' of various types of 'geometrical elements' derivable by 'extensive abstraction'; but excluding the introduction of more special properties by which straight lines are definable and measurability thereby introduced" (PR 96f.)

In distinction from the hierarchy involved in a compound individual, this is a hierarchy of abstractions, not of actualities. The point of this hierarchy is not that there are actual occasions anywhere that are characterized only by extensive connection without any more special characteristics. Some more specialized characteristics, such as those specifying ours as a geometric and then an electromagnetic epoch, are necessary. The point is that the less general features are contingent. Whitehead suggests that the most general level of characteristics, that of pure extensiveness, is probably metaphysical and applies to all possible worlds. (Hence this feature of our world would be parallel to God, Creativity, Eternal Forms, and Creaturehood as such in being an eternally necessary feature of reality.) But there could be creatures whose extensive connectedness was such as to make a different type of geometry important, or even such as to make geometry unimportant altogether. These features of our particular world are contingent. They are obviously compatible with the most general feature of extensive connection. In this sense one could say that the higher, more specialized forms of order were implicit in the deeper levels, but this language would again be misleading. For the more specialized forms of order were not necessitated by the more general. This is again due to the ubiquity of Creativity. Those actual occasions which, billions upon billions of years ago, began exemplifying those forms of order which now dominate our world were not following mere necessity. Some degree of self-determination (probably on the part of both God and the creatures) and hence contingency was involved. The world could have been otherwise. To speak of explication of implicate orders here would mute this contingency.

We *can* say that the deeper orders are implicit in the higher. The more general is implied by the less general. This shows the limitation of the Chinese-box analogy, for the smaller boxes can be removed from the larger without essential loss. But the characteristics of our cosmic epoch, with its laws and dominant family of straight lines, simply would not exist apart from the features making geometry in general important; and these geometric relationships in turn imply extensive relatedness in general. So the more particular implies the more general, i.e., the more general is implicit in the more particular.

If we are going to use the language of implicate and explicate in regard to *this* hierarchy, we would have to say that the more general levels of order were an explication of the more special ones (since they are implied by the more special ones). But no one would want to say this. And since we cannot appropriately say the opposite—that the more specialized are mere explications of the more general—it turns out that the implicate-explicate scheme does not apply at all to this hierarchy of abstraction, just as it failed to apply to the hierarchy of actuality.

Again we see that the key reason is the hypothesis that Creativity is the ultimate reality, embodied in all actualities. If Bohm accepts this notion, the implicate-explicate distinction will have to be disassociated from hierarchical notions. There *are* hierarchical features of the world and there *are* features of the world illumined by the implicate-explicate distinction, but these are different features.

9.4. Conclusion

The status of freedom, causality, and time are in the same boat. The denial of one implies, finally, the denial of the others. If time is unreal, in the sense that from an ultimate perspective there is no distinction between past, present, and future, then there can be no freedom, in the sense of self-determination in the moment. For if what still seems undetermined and hence future "right now" is in reality already as determinate as the past, then my feeling that I am deciding something in my apparently self-determining activity is an illusion. Likewise, if time is unreal, then there is no real causation, as distinct from logical implication. If the relations of the present with the past and the future are symmetrical, then the present implies the future as fully as it implies the past. Hence we can say that we cause the future only in the same sense as we cause the past, and this empties the word of any meaning.

Likewise, if there is no freedom, then time is ultimately unreal. (Again, I speak of the ultimate reality of time, not to reify it as another actuality alongside events, but as shorthand for the ultimate validity of the distinction between past, present, and future.) For if what happens a minute from now, a year from now, a million years from now, was already implicit in the world at this moment, with no genuine alternatives, then time is an illusion. The distinction between past, present, and future is merely an illusion: every event eternally exists, fully determinate. Time is invention or it is nothing, as Bergson said. So if there is no freedom, time is nothing. Similar reflections would show the dependence of causality upon time and likewise of time and freedom upon causation.

I have suggested that Bohm could formulate his intuition of wholeness without contradicting our deepest intuitions about time, freedom, and causation if he would highlight certain themes already present in his writings and drop others. In particular, the insight that the basic reality is a holomovement, or what Whitehead calls Creativity, should be strengthened, to stress that every event of enfolding and unfolding embodies this self-creative dynamism. Second, God should not be equated with this dynamic activity but regarded as the all-inclusive, intelligent, compassionate embodiment of it. Third, the distinction between implicate and explicate should be limited primarily to (1) the distinction between an event in itself (as subject) and an event for others (as causal superject) and (2) the distinction between the actual world with its enfolding and unfolding and the world as perceived through sensory experience. It can also be helpfully used for the distinction between unconscious and conscious experience, *if* it is clarified that the conscious aspect of experience is not merely an epiphenomenal explication of the unconscious depths. Fourth, the hierarchical features of reality should be stressed, but not in terms of the implicate-explicate distinction, since this application would contradict the ultimacy of creativity or holomovement and thereby threaten our deep convictions about time, freedom, and causation.

Abbreviations of Bohm's Works

RS   "Response to Schindler's Critique of My *Wholeness and the Implicate Order,*" *International Philosophical Quarterly* 20, 4 (December 1982):329–39
RV   "Conversations between Rupert Sheldrake, Renée Weber, David Bohm," *ReVISION* 5, 2 (Fall 1982)
S   "A Series of Talks Given at Syracuse University, September, 1982" (Unpublished manuscript)
W   *Wholeness and the Implicate Order* (London: Routledge & Kegan Paul, 1980)

Abbreviations of Whitehead's Works

PR   *Process and Reality,* Corrected edition, ed. David Ray Griffin and Donald W. Sherburne (New York: Free Press, 1978)
RM   *Religion in the Making* (New York: Meridian Books, 1960)

10.   John B. Cobb, Jr.

BOHM AND TIME

When reading David Bohm I am chiefly delighted by insights and daring speculations that have the ring of verisimilitude. But from time to time I am jarred by ideas that I cannot appropriate and that raise doubts in my mind as to whether I have been correctly interpreting other passages. I have decided, perhaps more for the sake of my own peace of mind than in objective response to what he has said, to think in terms of two Bohms. I would like to speak of them as the earlier and the later Bohm, since that is the fashionable way to deal with these matters, and indeed I do think one set of images was more prominent in earlier work and another set in more recent papers. But what I am for convenience calling "Bohm I" intrudes into his most recent writings as well. The differences between the two sets of images, which of course also overlap, can be focused on his statements about time.

In section 10.1, I will lay out the two sets of images and the concepts that explicate what is implicit in them. In section 10.2, I will suggest a further development of "Bohm II" as this bears on time. Much of what I propose there comes from Whitehead, but for the most part I will not present it as a discussion of Whitehead's ideas *per se*. In section 10.3, I will discuss what Bohm has to say about transmissions of information that are not limited to the speed of light. I believe that in part his explanation here reflects Bohm I, whereas I think it could be more satisfactorily developed in terms of Bohm II.

Before entering into the substance of the discussion I need to confess my discomfort and hesitation. I cannot even read the mathematics at the heart of the discussion among scientists. This means that when I suggest how physical phenomena are to be interpreted I do so from outside science and without any clarity as to what my suggestions would mean in technical formulations. I proceed to make them out of a conviction, shared with Bohm, that better concepts are needed to interpret the mathematics and to relate the mathematics to future theoretical development as well as to our total experience of the world. But it is a very questionable enterprise to make such proposals when one does not understand the mathematics at all!

10.1. Earlier and Later Sets of Images

For many years Bohm has felt keenly the inadequacy of a purely phenomenal and mathematical account of physics. He has sought an interpretation of the phenomena that would explain them in terms of an underlying reality and thereby show that the physical world does not depend on human observation for its existence. I am convinced of the great importance of this enterprise, if not for physics, then for life in general and especially for religion. Bohm, too, sees this connection to religion as the work of binding things together. But he is also convinced, and this corresponds to my intuitions, that physics itself suffers from its acquiescence to phenomenalism. It is my hope that he will be able to demonstrate this in due course, for I realize, as he does, that until his realistic models generate new predictions that can be empirically tested, many physicists will pay little attention.

Often the quest for a realistic model of the world is associated with the desire to re-affirm the mechanistic vision. That vision took such firm hold on the Western mind that it has been difficult to imagine other ways of conceptualizing a real world. The major reason for limiting science to the phenomena, first in Kant's philosophy and more recently among leading physicists, has been to prevent science's mechanical model from being applied to reality. Kant's point was a purely philosophical one. He assumed that mechanical models worked for the phenomena. But subsequently it turned out that they worked only by being applied in paradoxical fashion. Both particle and wave are models derived from mechanical concepts.

Those who have sought behind this puzzling duality another kind of entity explanatory of both—hidden variables, for example—have generally pictured these as well in images derived from the mechanical world view. Bohm, on the other hand, accepted the breakdown of the mechanical world view as irreversible. He sought for a different mode of causality and a different image of reality. And here, too, I find myself in wholehearted agreement. Both Bohm I and Bohm II are nonmechanistic realists, but the ways they image a nonmechanistic reality differ.

Bohm I shared with mechanistic realists an interest in hidden variables. This meant for him that beneath the level at which indeterminacy reigned there must be another level in which the physical world was determinate, that is, a definite situation existed whether human beings observed it or not. But for him the hidden variable would not be some kind of mechanistic particle. Instead it would be a deeper order related to quantum phenomena in a way analogous to that in which the level of quantum relates to the macroscopic world. The causal relations that interested him were between one level and the next, rather than those among the entities at one level.

Bohm I could use quite deterministic language. He thought of the phenomena at one level as fully determined by the reality at the deeper level. Since he did not want to draw deterministic conclusions with respect to human freedom, he proposed that these could be avoided by positing an infinity of levels. But fundamentally he seemed little disturbed by determinism as long as it was not mechanical in character.

The relations among the levels could be formulated in terms of implicate and explicate orders. What is explicate at one level is implicate in the underlying level that explains it and gives rise to it. This underlying level in turn is explicate in relation to a still deeper level. It is this sequence of levels which, he suggested, might be infinite, although he did not insist on this point.

With respect to time, Bohm I could suggest that it might function at some levels but not at others. In particular, he made the suggestion that, in the deeper implicate levels, time would be implicit rather than explicit. This could mean that past, present, and future appear in the explicate order as separate and distinguishable, whereas in an implicate order they are all together as one. This could explain those rare phenomena that seem to transcend or violate the normal temporal sequence. This speculation has a long and respectable history and seems to gain some support from those aspects of physics that view the course of events as in principle reversible.

Bohm has demonstrated the fruitfulness of the image of implicate and explicate orders in many fields. For example, it is suggestive with respect to the relation of language to speech and of unconscious experience to conscious. It parallels the profound studies of Polanyi on tacit knowledge.

Nevertheless, I do not find this imagery adequate, and I have rejoiced to find it developed and transformed into something quite different in Bohm II. I suggest that this shift can be viewed fruitfully in terms of the language of "enfolding" and "unfolding," which he has used in close conjunction with "implicate" and "explicate." The chief difference is that enfolding and unfolding suggest a reciprocity that is not evident in the successive layers of implicitness. That is, when the imagery is of implicate and explicate orders, it is not evident that the explicate order has any autonomous effect upon the implicate order that gave rise to it. The causal relationship seems clearly to be in one direction, from the implicate to the explicate. But if the implicate order enfolds the explicate order and then unfolds into the new explicate order, then the causal relation moves in both directions. Bohm II can think of the way the implicate order gives rise to the explicate one as leaving something, however little, for the explicate entities to decide. The temporality of the explicate order then necessarily affects the implicate order. Bohm II has no need for an infinity of levels to avoid the denial of self-determination in human experience or elsewhere.

Bohm II places a central emphasis on internal relations. First, every entity in the explicate order fragmentarily enfolds, or is internally related to, the whole. This whole is the holomovement that also enfolds every entity. Furthermore, the entities enfold one another. Bohm is clear that the fundamental feature of the mechanistic world view that must be overcome is the idea of parts that are separate from one another, that only additively constitute the whole. Bohm is convinced that the evidence of physics indicates that instead there is an undivided wholeness constituted by the way in which everything enfolds the whole.

Although Bohm still at times suggests that the "wholes" in question may constitute an infinite series, when he is following the logic of Bohm II he treats the whole or the holomovement more univocally. No doubt there is much of which we are yet quite ignorant, and some of this may be well conceived as "deeper levels." However that may be, all that we do not know together with all that we do know is part of one holomovement. The idea of an infinity of holomovements simply confuses this picture.

The idea of the hidden variable does not fit with Bohm II. In its most common use a "hidden variable" was an entity or field more fundamental than those to which our present equipment gives us access and whose behavior explains coherently what is paradoxical at the level now treated by quantum theory. Bohm's idea of a hidden variable did not quite fit this pattern, but an implicate order determinative of the explicate one had sufficient analogies for the term to be useful. If, however, we think in terms of enfoldment of the whole into each part and of all the parts into the whole, there is no need for a hidden variable. Of course, much is "hidden" since internal relations are not observable in the same sense as external relations. But this is just to say that what is observable is never reality, it is always phenomena or appearances. To make that point we do not need to speak of a hidden variable. Indeed, such language is seriously misleading.

10.2. Time and the Later Bohm

It is my belief that if Bohm II can be more fully separated from Bohm I and can be developed further along lines that he has himself suggested, a more adequate and powerful conceptuality will emerge. One result would be a much more consistent recognition that the "particles" of which Bohm continues to speak are constituted by successive enfoldings and unfoldings. They are not, then, real entities that move continuously through a given space but successive acts of self-constitution out of the whole and all its subwholes or parts. These acts are occurrences, happenings, or events rather than the sorts of things that endure through time. A so-called particle is a succession of such events in which later members appropriate both energy and form from their predecessors. A wave is the effect of any one of these events upon successor events that do not appropriate its total form so fully. There are not two types of reality, particles and waves, but one undivided wholeness in which some of the emerging patterns lead us to speak of "particles" and others of "quantum potentials" or of waves.

Since there can be no event that is not an enfolding of its world, Bohm's emphasis on internal relations is fully sustained. Indeed, the deepest meditation on internal relations must lead to the primacy of events over particles and quantum potentials or waves. Enfolding is, can only be, an act or an event. The image of particles enfolding one another does not work when particles are thought of as something other than a sequence of events with very similar forms.

I believe that Bohm understands and agrees with what I have said here. My complaint is not that he has not said it; he has indeed explained that so-called particles are really "world tubes." My complaint is that he has not yet made full use of this conceptuality in his physics. Even his most recent papers speak in terms of particles and of quantum potentials as if these were the real entities that made up the world. There is still a tendency to seek the resolution of such dualisms at a "deeper" level. I think his own work points clearly to the fact that this "deeper" level is the events themselves, but again he has not stated this consistently.

The enfolding is the self-constitution of the event out of its world. The unfolding is the way in which the completed event contributes itself to the whole and to the future. We humans are most aware of this as sensory form, but of course in the world of quanta the effects of one event on another would not have sensory elements. We can get a better analogy from our own experience if we think instead of the flow of experience as itself a succession of enfoldings and unfoldings. In each moment my experience enfolds the whole. Of course this is *very* selective. Most of what is enfolded comes through the body and is mediated to my experience through the events occurring in my brain. But there is also the enfoldment of my previous experience. In one moment most of what I was thinking and feeling in the previous moment continues or is renewed. It may have a large sensory component, but my present relation to my past experience is not mediated by my sense organs. The previous momentary experience constituted itself by enfolding its world and then unfolded itself for my present momentary experience to enfold. In addition, of course, its unfolding affected events in my brain and in the wider world as well as the whole.

Bohm sees clearly that since both quantum occurrences and human experiences are acts of enfolding and unfolding, his vision overcomes the dualism of mind and matter. Yet he continues to use the language of "mind" and "matter" as he does that of "particle" and "wave" as if they were real. This encourages him to think of overcoming dualism through appeal to a deeper level, as if mind could act on matter only through its enfoldment in the whole, which then unfolded into the matter. I do not think he believes this, but my complaint is that his rhetoric is not free from this implication. If he would consistently speak of events of conscious human experience on the one side and events in the physical world on the other and would pursue the similarities to the end, the enfolding by human experience of physical events such as those occurring in the brain cells and the enfolding of human experiences by these in turn would cease to be especially puzzling. Again the indivisible wholeness of events is the underlying reality that brings mind and matter into the one holomovement.

This modification of Bohm's rhetoric needs to be more fully extended. Bohm I thought of the causal relation only between the implicate order and the explicate one. Bohm II recognizes that the entities (events) of the world enfold one another directly as well as *via* the whole. Nevertheless, the continuing influence of Bohm I prevents him from accepting this point wholeheartedly. He wants to affirm that the fullest instance of enfolding is the enfolding of the whole and that the enfolding of one event by another is less real or less important. If Bohm II could be freed from this lingering prejudice derived from Bohm I, the enfolding by each event of all the other events in the world could be recognized as fully real and important. This would not entail any denial that each event also enfolds the whole and that apart from this enfolding there would be no event at all. One would not have to suppose that the sheer occurrence of the event or the order it manifests can be fully derived from its enfolding of past events. There is an absolutely necessary role for the whole to play in relation to every event. Hence my criticism of Bohm is not radical. My complaint is simply that he has not fully acknowledged that the enfolding of other worldly events is fully real and also absolutely necessary to the constitution of any worldly event. Surely the primary way in which my present experience is related to my just preceding experience is immediate, that is, not mediated by the whole. This is not to deny that the exact way that past experience is enfolded into the present one is affected by all the other enfoldments that occur and especially the enfoldment of the whole.

What has all this to do with the issue of time? A great deal, I think. If the world is made up of events largely constituted by their enfoldment of their worlds, then the notion of reversibility simply does not have any possible application to the fundamental units of the real world. I think this point is of sufficient importance to emphasize by contrast. If the world is made up of entities that are related to one another externally, so that their spatial configurations change without altering their fundamental nature, then it is an open question as to whether certain configurations can be repeated in the course of time. One may even speak of passing through a series of configurations in reverse order. If time is correlated with movements of this sort, then it is possible to speak of time running backwards. I think there are problems with this way of thinking even in relation to a world of particles in motion, but the discussion of such possibilities is understandable. Such profound thinkers as Nietzsche have derived from this world view the notion of a cyclical universe in which every situation would eventually repeat itself again and again.

If the world is composed of events that enfold their past worlds, then none of this makes sense. We cannot exclude the possibility of certain *abstract patterns* repeating themselves for observation. But what is happening *concretely* is in fact never a repetition of what has happened before. The world that is enfolded by two events is never exactly the same.

This lack of identity of the world of two events can be demonstrated easily if we take the light cone as fundamental for enfolment. Every pair of points defines two distinct light cones. Of course there will always be overlap, and, in the case of contiguous contemporary events, the overlap will be overwhelmingly large and the differences vanishingly small. But if one event is in the light cone of the other, the difference will be quite significant, since one is unfolded for the other but their is no reciprocity. There can obviously be no repetition of an event, since the new event will have enfolded the old one along with other events that the old one could not have enfolded.

There is increasing evidence now that the light cone does not have the absolute character long attributed to it. Bohm believes there is instantaneous transmission of information, to which I will return in section 10.3. I mention it now only to indicate that nothing Bohm states in his discussion of this issue undercuts the notion that the light cone continues to be an important physical reality. That is sufficient to establish the nonrepeatability of events.

The nonrepeatability of events in turn establishes the irreversibility of the actual process. Time, *as a function of this process,* is real for all observers. It characterizes the holomovement as much as any subordinate process within it.

Nevertheless, time is not in itself a fundamental reality. It is an abstraction from the process. Since it is an abstraction illustrated in *every* succession of events, and since a world of events is *always* a world of successions, time is *necessary.* But we must not commit the fallacy of misplaced concreteness by reifying time. What really exists is a succession of events. They are by their very nature related to one another as past and future or as contemporary. These relationships are temporal. But there is not something to be called time *apart from* these actual relations.

There is another respect in which recognition of the priority of events displays the crucial and inescapable nature of time. The units of which the world is made up have a temporal dimension or aspect. This, of course, is not the case with "particles," as conceived thus far by physics. The idea of the particle is that of an entity that moves through time but is fully what it is at any instant. In other words, its essential nature is not affected by time. It is inherently nontemporal. The same entity might have been at some other time. It just happens to exist at *this* time. Time must then be thought to be either (1) an independent reality within which the particle is located or (2) a function of changing positions of particles relative to one another.

An actual event, in contrast, is four-dimensional. This is obviously true of what are usually called events, such as presidential elections or conversations. But even when we analyze these large-scale events into the component events of which they are made up, all four dimensions remain. A moment of human experience, for example, takes place with spatial location but also with temporal location. Also, the spatial locus may approximate a point in space, but never becomes one, and the temporal locus may approximate an instant, but it never becomes one. For something to happen, such as a quantum event, four dimensions are required.

This does not mean that *within* a single unit event there is a before, a contemporary, and an after. These are relations that exist *between* unit events, not within them. It does mean that the unfoldments whereby events become effective for other events do not constitute a continuum.

Successive unfoldments are separated from one another by a finite interval that can be analyzed in both temporal and spatial terms. Since movement in the sense of *locomotion* is the difference of location of successive unfoldments within an inertial system, it necessarily involves discontinuity, however negligible.

Time and motion, as these are ordinarily understood, belong to the explicate order. If we consider the process of enfoldment and unfoldment itself as belonging to the implicate order, then we can understand that there is a sense in which, after all, time is implicit in the implicate order. The act of enfolding and unfolding cannot be divided into subacts that are temporally successive. It is a unitary act that either occurs in its entirety or does not occur at all. Some meditational disciplines concentrate on the full realization that at any moment one is just such an undivided act, which is not temporal in the sense in which time refers to the successive unfoldments. If the meaning of time is derived from the explicate order, then there is an important sense in which the implicate order, that is, the act of enfoldment and unfoldment, is timeless. Nevertheless, this timeless act is an enfoldment of other acts which constitute its past and it unfolds itself into other acts which constitute its future. Time is a function of the relations of these events.

10.3. Instantaneous Transmission of Information

Some of the most interesting of Bohm's speculations about time have to do with phenomena that indicate that the speed of light is not as absolute as Einstein supposed in determination of causal influence. Bell's theorem has been particularly important, since it indicates that events that are causally independent, in terms of the Einsteinian light cone, do affect one another. The evidence is that between noncontiguous events there is, in addition to mediated influence, another type that is immediate.

If we accept Einstein's view that no signal can be transmitted through space faster than the speed of light, then we must hypothesize that spatially separate events may influence one another apart from any transmission of signals through the intervening space. That is, the events between the two communicating events need not be involved. One event may enfold another event at some distance as soon as that event has unfolded.

Bohm suggests that what is communicated in this immediate way is not energy but only information. This conforms remarkably to Whitehead's speculation on the same topic. Whitehead emphasized that every event in the light cone of a concrescing event is "prehended." That was his word for "enfolded." This prehension would be "purely physical" in the sense of taking account of all those events in terms of their own physical (i.e., energetic) character. He was open philosophically to the possibility that the pure physical prehension of noncontiguous events might be immediate, but he said that, as long as physical science adopts the view that there is no action at a distance, one should regard this possibility as negligible or nonexistent.

However, Whitehead believed that every event has some element of self-organization that transcends its purely physical or energetic character. Every event embodies in itself some *form*. The enfoldment of this form of another event, separate from its physical energy, is what Whitehead called "hybrid physical prehension." (It's still "physical" in that it is a prehension of another event, another actuality.) Whitehead believed that hybrid prehensions would not require mediation through spatiotemporally intervening events. He regarded mental telepathy as an example of this relation. He did not know, of course, of Bell's theorem, but it is my impression that he would have been delighted by it.

Bohm comes to a very similar conclusion in meditating on Bell's theorem. He sees no transmission of energy between simultaneous events. What is communicated is information. The close agreement of Bohm's thesis and Whitehead's proves nothing about their correctness, of course, but I do find it remarkable considering the half-century that intervenes between them and the virtual absence of such speculation in that interim.

This doctrine opens up a whole field of speculative possibilities. If information can be communicated without mediation between events that are precluded from mediated communication, can it also be immediately communicated by one event to another even though there are also mediated communications? To put this another way, can a present event receive from past events information that has not been mediated by spatiotemporally intervening events? Can memories, for example, have this character? Does the "morphogenetic field" as developed by Rupert Shelldrake suggest this kind of immediate communication of information? My own inclination is to answer yes to these questions. If so, we will have a far richer sense of time than that bequeathed to us by Einstein, while retaining the principle that there is a limit to the speed with which physical energy can be communicated.

Also, if information can be communicated between some contemporaries, in Whitehead's sense (that is, events between which there can be no communication at the speed of light), should we assume that some communication exists between all contemporaries? Are the events for whose enfoldment we have evidence simply ones that have particularly marked efficacy? Or is the enfoldment of contemporary events highly selective, occurring only when commonality of pattern or common history attunes one event to the other in a peculiar way?

Bohm develops his speculation in a way that is promising for answering this question. If I understand him correctly, he thinks that, since the physical features on which special relativity theory is based do not affect instantaneous communication between spatially remote events, this communication would be instantaneous in terms of a true or cosmic time from which particular inertial systems vary. There is a time for the whole or the holomovement as well as time as measured by diverse inertial systems. And it is instantaneity in this cosmic time that governs the transmission of information between the events described by Bell's theorem.

The question I have raised is thereby partly answered. In addition to influence mediated physically at the speed of light, there is also immediate influence from events that have just occurred in cosmic time. There would be no influence from events in the cosmic future, and there would presumably be no influence from other contemporary events except as mediated by those that have just occurred in cosmic time. This leaves unanswered the question as to whether there is some transmission from all events that have just occurred in cosmic time or only from special events linked in specifiable ways. Bohm's answer seems to move in the former direction, as when he says that "the interaction of each pair of particles may then depend non-locally on all the others, no matter how far away they may be."

I hope that the reinstitution of cosmic time can be sustained. It has seemed to me that general relativity theory does require a cosmic frame of reference that relativizes those of particular inertial systems. Philosophers have confused matters when they have treated special relativity as providing a metaphysical basis for rejecting a cosmic present.

Nevertheless, I find in Bohm's recent discussion of this topic a remnant of Bohm I. This is his reintroduction of the "ether." True, he puts quotes around the word to indicate that he does not mean quite what the word has been taken to mean. But his discussion suggests that, despite the acknowledgment of the importance of events at some places in his writings, he has not rid himself of the sense that events presuppose something more substantial. Because I believe this to be the crux of my quarrel with Bohm, I will develop the point in some detail.

In his recent article on "Measurement Understood through the Quantum Potential Approach" (with B. Hiley), Bohm introduces the notion of the ether as follows: "This theory implies that even in the vacuum, there is an actual random fluctuation of the field, associated with what is commonly called the zero point energy. This is reminiscent of Dirac's 'ether.' What are called particles are then actually states of generally conserved forms of excitation of the 'ether.' . . . In view of the non-local connections, we are forced to assume that this 'ether' constitutes a special frame."

This passage is one of many in which Bohm recognizes that particles do not have the character usually assigned them, although I have already indicated that he himself seems to keep this point in mind less consistently than I would like. But instead of viewing the particles as a succession of enfoldings and unfoldings, as he sometimes does, he reverts to the notion that there must be something underlying, an "ether," that is subject to excitation. He acknowledges that this cannot be detected by ordinary means, and I do not think it plays any positive role in his theory. It simply expresses his continuing tendency to posit something behind the events—behind the excitations, as he calls them here.

My conviction is that his formulations in this paper would be clearer and more convincing if he renounced the ether. The point is that there are events occurring in the vacuum. But events should not be thought of as occurring *to* some underlying substance or subject. If we must speak of an ether, let us speak of an "ether of events." If that does not suffice, we could note that every event is an instance of the universal process of enfoldment and unfoldment. Whitehead called this "creativity." Creativity is the ultimate reality, but it is not a substance to which things happen; it *is* the happening. When we find that events have immediate, simultaneous efficacy despite their spatial separation, we have learned something about the nature of events and their enfolding and unfolding. We do not need to posit an ether.

My point here is similar to my criticism of the appearance in Bohm's paper of the positing of particles and quantum potentials as separate types of reality. Here we find him adding the ether as a third. Why multiply types of reality when one will do? What we need are events and their modes of enfolding and unfolding. I indicated above how different modes of enfolding could lead to particle and wave phenomena. Now we learn that in addition to locally transmitted influences there is a transmission of influences that is not mediated through intervening events. Why not simply recognize that events enfold forms from distant events immediately?

The only advantage I can see in the multiplication of entities is that it might seem to be intuitively clearer. Perhaps physicists are so fully enslaved to the mechanical world view that only concepts to which that world view gave rise are felt to be intelligible. Then I suppose we must have particles and quantum potentials and ether. But surely what we need, in full agreement with what Bohm himself argues elsewhere in great detail, is a different way of thinking. I propose, based on suggestions of Bohm, that it is possible to think of events and sequences of events and to find such thinking fully intelligible. This is a realistic and nonmechanistic way of conceiving the world. It seems to fulfil Bohm's intentions and to be compatible with most of his concrete proposals. I earnestly hope that substantialist habits of mind can be overcome and that physics can reconstitute its formulations on the basis of event thinking. I believe that could lead to going beyond the dualisms by which it is now plagued.

11. Ian G. Barbour

BOHM AND PROCESS PHILOSOPHY:

A RESPONSE TO GRIFFIN AND COBB

The papers by John Cobb and David Griffin provide detailed and illuminating comparisons of the conceptual schemes of Bohm and Whitehead. The similarities are indeed striking. However, I see greater differences between Bohm and process thought than these papers do, especially in relation to three crucial topics: timelessness, causality, and nonlocality.

11.1. Timelessness

Cobb cites Bohm's assertion that "past, present and future are all together as one in the implicate order." Cobb also mentions that in meditation there can be a timeless and undivided awareness. Bohm's paper maintains that as consciousness turns to deeper unconscious levels the moments become more similar; at a very deep level, "all 'nows' would not only be similar—they would all be one and essentially the same. So one could say that in its inward depths, now is eternity." Through such experiences, Bohm suggests, we have personal contact with the timeless implicate order.

Persons in many religious traditions have indeed described mystical experience as timeless. But how is this to be interpreted? The intense experience of unity with all things seems to transcend all boundaries of space and time. Meditation may achieve a level of consciousness in which the normal flow of thought and the shifting patterns of attention cease, and one is not aware of the passage of time. There is also a liberation from bondage to time and the transitory world of impermanence and flux. Such an enduring, all-encompassing relationship beyond the vicissitudes of time can be described as participation in eternity.

Beyond the experiential meanings of timelessness, many religious traditions have characterized the divine as timeless (for example, the imperishable, impersonal Absolute of Vedantic Hinduism or the eternal, unchanging God of medieval Christianity). But neither the biblical God nor the God portrayed by Whitehead can be described as timeless. To be sure, Whitehead speaks of God's "primordial nature," his unchanging purposes and envisagement of eternal forms, but this is an abstraction from the actuality of God that includes his "consequent nature" in temporal interaction with the world. In both religious and metaphysical functions, Whitehead's God, like the biblical God, is an active agent in an ongoing historical process. Hartshorne's writings have explored the differences between timeless and temporal views of the divine in both Eastern and Western thought.[1] In this religious dimension, I see a real divergence between Bohm and Whitehead.

From the scientific side, the theory of relativity has sometimes been invoked in support of the idea of timelessness. Bohm's paper states that in relativity theory "the whole of space-time has to be considered as a block." The phrase "as a block" suggests that the four-dimensional continuum is to be visualized by the spatialization of time. But one could just as well refer to the temporalization of space. As Čapek points out, in relativity theory there are some events that are past for one observer and future for another observer, but for any two events that could be causally related there is an absolute distinction of past and future for all possible observers.[2] There is no way in which a future event could influence the past or the present, according to relativity theory. It is obscure to me in what sense Bohm thinks of past, present, and future as "together" in the implicate order. Far from being timeless, the world as understood by modern physics is a dynamic and ever-changing spatiotemporal pattern. For Whitehead, then, temporal process characterizes all levels of reality, whereas Bohm speaks of temporal and timeless levels in the structure of reality.

11.2. Causation

Cobb and Griffin do note some differences between Bohm and Whitehead here. Cobb points out that for Whitehead events enfold each other directly, whereas for Bohm causality is mediated *via* the whole. Griffin argues that Bohm excludes horizontal causation and the direct grasping of one entity by another and instead portrays vertical causation ("projection from the bottom up and reinjection from the top down"). Griffin replies that if entities interact only *via* the whole it is difficult to account for human freedom, our common experience of direct interaction, and the rarity of nonlocal effects. Bohm's paper refers to a horizontal implicate order in addition to a vertical implicate order, but his detailed elaboration deals only with the latter. Bohm's scheme tends to be monistic, whereas Whitehead's is fundamentally pluralistic, with a multiplicity of centers of activity. As a consequence, Bohm has to struggle continually to find room for human freedom, whereas for

Whitehead it is an extension of the self-determination that characterizes all entities.

Yet Griffin also portrays similarities between Bohm and Whitehead and says that they agree in holding that there is *no direct causation* between explicate objects. For Whitehead, he says, the line of causation passes through an event's subjectivity, which is hidden from view and might therefore be considered implicate. I would reply that for Whitehead subjectivity is a moment of self-determination in isolation from the whole, whereas in Bohm's implicate order individuality is absorbed in the whole and discrete self-determination appears minimal. Despite Bohm's concern for creativity, his analogies for the implicate order are usually *deterministic*. As one analogy, he describes an experiment in which a drop of ink is placed in a liquid between two concentric cylinders, one of which is rotated, spreading the drop into a complex spatial pattern. If the cylinder's motion is reversed, the drop will reappear determinately from its implicate form. In the case of a hologram the image of the whole can be obtained from any part because the light waves came initially from objects with fixed spatial relationships. Bohm refers to the projection of the explicate order from the timeless implicate order. "Projection" in this context seems to be a geometric analogy. "Implication" is a timeless logical relationship, as Capek points out. It suggests a deterministic pattern in which creativity and freedom have to be introduced as exceptions rather than as fundamental components.

In short, I see many parallels between Bohm and Whitehead, and I am glad that Cobb and Griffin have welcomed Bohm to the process club. It is also impressive that Bohm has recently made considerable use of Whiteheadian categories. But I see some basic points of difference that are hard to reconcile. I am quite willing to accept paradoxes as long as they are grounded in experience, and as a physicist I have learned to live with complementarity. I have defended the use of pluralistic conceptual models that cannot be neatly integrated into one comprehensive model.[3] I am particularly sympathetic with Bohm's conviction that we need to draw from both East and West and from both meditational and scientific disciplines. But I wonder if one doesn't finally have to come down on the side of a basically monistic vision or a basically pluralistic vision of reality.

11.3. Nonlocality

Cobb and Griffin both cite the Bell's theorem experiments as evidence that information can be communicated instantaneously. They assert that claims for mental telepathy are more plausible in the light of these experiments. Cobb says "this could explain those rare phenomena that seem to transcend or violate the normal temporal sequence." Let us consider these experiments and their interpretation.

A photon source gives off two photons in opposite directions (to the left and to the right, let us say) with opposite but randomly oriented polarizations (one up, the other down, for example). In the path of the left photon is a filter that allows only photons with a particular orientation to pass through to strike detector *A*. A similar filter and detector *(B)* are in the path of the photon moving to the right. The record of the two detectors might look like this:

$$A: 1 \ 0 \ 1 \ 0 \ 0 \text{ etc.}$$
$$B: 1 \ 0 \ 0 \ 1 \ 1 \text{ etc.}$$

Here the first pair of photons both registered (1). Of the second pair, neither one registered (0). Of the third pair, *A* registered but not *B*, and so forth. Of these five photon pairs only the first two (40 percent) show *A* and *B* correlated. If the angle between the two polarizing filters is changed, the fraction of correlated tallies will change.

In recent experiments, the orientation of the polarizing filter at *A* was changed while the photons were in flight—too late for a signal traveling at the speed of light to reach *B*, several meters distant, before the photon arrived there. It might appear that one could use the device to communicate instantaneously, from *A* to *B*, but this is not the case. The photon pairs emerge from the source with random (though correlated) orientations. The record at *B*, taken alone, appears completely random—whether the polarizer at *A* is switched or not. The change in the statistical correlation between *A* and *B* can be detected only when the records from *A* and *B* are brought together. The information is in the correlation, not in *B*'s record alone, and the correlation is known only when the record of detector *A* is independently transmitted to *B*.[4] The experiment thus does not seem to make more plausible the claims for mental telepathy—much less the claims of some authors that the experiments show the possibility of precognition of future events (which, I would argue, is inconsistent with the irreversibility of time and with the presence of indeterminacy and freedom in the world).

There is indeed a nonlocal correlation between events at *A* and at *B*, but it exists only because of a common initial event at the source of the photons. In quantum mechanics, the two photons traveling in opposite directions must be described by a single wave function that extends to the two detectors. It cannot be assumed that polarization is a property of each particle separately, as would be the case in classical physics. This does not mean that we have to abandon realism and treat quantum theory merely as a calculating device for predicting observations, as positivists hold. Nor should we conclude that reality is created by the human mind and has no existence independent of consciousness (as idealists hold). But we can be realists only if we recognize that the reality of the microworld is very different from the reality of the world of everyday objects and that it cannot be described by classical concepts. The quantum state is the state of a total system, which in this case includes both photons; the measurements are a product of the interaction of that system with the instruments of observation.[5] In the Bell's theorem experiments we can see a vindication of Whitehead's rejection of simple location (the idea that all entities exist at determinate locations in space and time). As Whitehead insisted, we always deal with interactions and not with separate substances possessing inherent properties. We do have to treat the system as an indivisible whole. Only if we imagine separate particles do we have to imagine causal connections transmitted faster than the speed of light.

I would thus want to affirm the wholistic thrust of Bohm's thought, which is consistent with the necessity of treating the wave function as a whole. But the "instantaneous collapse" of the spatially distributed wave function is not really a form of instantaneous communication. Whitehead does speculate that action at a distance, if confirmed, would not be inconsistent with his theory of hybrid physical prehensions, but his attack on simple location is more fundamental to his whole system. In sum, I do not find support in either contemporary physics or process philosophy for ideas of timelessness, "vertical causation," or instantaneous communication.

Notes

1. Charles Hartshorne and William L. Reese, *Philosophers Speak of God* (Chicago: University of Chicago Press, 1953).

2. Milič Čapek, "Relativity and the Status of Becoming," *Foundations of Physics* 5, 4 (December 1975):607–17.

3. Ian Barbour, *Myths, Models and Paradigms* (New York: Harper and Row, 1974), chap. 5.

4. Henry Stapp, "Quantum Mechanics, Local Causality, and Process Philosophy," *Process Studies* 7, 3 (Fall 1977):173–82; Heinz Pagels, *The Cosmic Code* (New York: Bantam Books, 1983), chaps. 12, 13; see also Bernard d'Espagnet, *In Search of Reality* (New York: Springer-Verlag, 1983).

5. Henry J. Folse, Jr., "Complementarity, Bell's Theorem, and the Framework of Process Metaphysics," *Process Studies* 11, 4 (Winter 1981):259–73.

12. David Bohm

First, let me express my appreciation of the papers of John Cobb and David Griffin, for their careful and serious consideration of my work. I feel that they have understood the essence of what I have been aiming for in this work. Their criticisms are cogent, and after careful consideration I feel that some of them are justified. A complete answer to all the issues that they raise would require a long paper, and I should therefore here touch on only a few key points.

To begin with, I would like to say that my ideas are meant only as proposals for inquiry. In such inquiry, one learns and changes one's notions of reality. My own notions have changed in this way, through interaction with those who attended the conference and through reflection on the various arguments that have been put forward. It seems to me that there can never be a final and totally consistent metaphysical account of reality. Rather, in the development of our theories, we are taking part in a continual process of learning, the unfoldment of insight. In this sense, I am proposing that Whitehead's notion of reality as process be extended to include even our own metaphysical notions about process itself. This brings us to a higher level of consistency, in which what we say and what we are actively doing as we say it are coming into closer harmony. Otherwise, we are caught up in the inconsistency of saying that all is in a process in which everything changes except for our basic ideas on process itself.

I shall begin by discussing Cobb's paper, in which he refers to "Bohm I" and "Bohm II." Perhaps I could now say that this paper is written by "Bohm III." I now regard much of my earlier work on hidden variables and quantum potentials as a simplification and abstraction from a more fundamental process, in which the "particles" of physics would be recurrent moments, unfolding from an implicate order and enfolding back into it. With this, I feel sure that both Cobb and Griffin would agree. And when I proposed an "ether" I did not mean to imply a space-filling substance (as is indicated by my use of quotation marks). Rather, I was mainly trying to give an image to what I have called the "zero point movement of fields in the vacuum." As pointed out in my paper, "Time, the Implicate Order, and Pre-Space" (TIP), within this movement there are all sorts of properties (e.g., vacuum polarization) that suggest that space is some kind of plenum of process, rather than sheer emptiness.

Modern quantum mechanical field theory indicates that there are at least two kinds of implicate order. The first of these is the set of physical fields themselves. The second is what I have called the "superimplicate order," which is represented mathematically by what may be described as the "wave function of the first set of fields." The superimplicate order organizes the first implicate order. For example, when the "quantum state" of a field is excited, the unfoldment of the field into momentary localized pulses is such as to produce on the average just the sort of recurrent structure that Cobb calls for as an explanation of what underlies the notion of a "particle" in modern physics. And it answers Griffin's question as to why these "particles" (along with other structures constituted out of them) are stable. However, in doing this, one has to take into account the fact that the pattern of recurrent pulses depends, at least in principle, on the quantum state of the whole universe.

As for the causal connections, one finds that these are of two types. There is one kind of connection that applies directly on each level. This answers Griffin's question as to why the recurrent pulses are *contiguous.* And there is another kind, in which the very nature of the causal connection within the first level of the implicate order depends on the superimplicate order. This is, of course, what is responsible for the *stability* of patterns of pulses. I feel that this is very important, as it indicates just how the overall organizations of the "particles," and indeed their very existence within some stable range of forms, depends on the state of a greater whole. It seems to me that similar relationships are to be found, for example, in societies of human beings, in which the stability of relationships between people depends crucially on the state of the whole of society. It thus appears to be plausible to suppose that there is a corresponding dependence of relationships in the nervous system, the brain, etc., on the state of the whole.

What is missing in all this, as Cobb and Griffin would point out (and also as I have long felt), is an account of how the state of the whole could also depend on the states of the parts (or subwholes). Clearly, in examples such as that of society, there *is* such a dependence. However, as quantum mechanics now stands, there is no such dependence of the superimplicate order on the first implicate order *in physics.* I think that it is fair to say that this is a defect of the idea. However, it is possible that this may be a key clue as to how the quantum theory itself should be extended, so as to include the possibility of a mutual dependence of the two levels of implicate order. So what is required is a further step beyond the present quantum theory. Since the conference, I have had some ideas on this point that could possibly show how this could be done, but it is too early to discuss them here.

I feel that the step indicated above would go a long way toward answering the criticisms of Cobb and Griffin. This point is carried further if one considers how in my paper (TIP) I showed that the relationship of moments in the first implicate order is similar to that implied by Whitehead's notions of concrescence, creativity, and superjection.

As soon as one thinks of these two levels of implicate order, one immediately asks: "Why not three or more?" Thus one comes to the notion of a "vertical dimension" to the implicate order. Since there is no reason why the number of levels should be limited, I came naturally then to the idea that the number may be infinite. The advantage of having an infinity of levels is that it provides an image of how any finite context is incomplete, and it leaves room for new and creative action originating beyond the context in question. And, of course, if the connections in the "vertical direction" go both ways, then this should help to answer the objections of Griffin, that the finite and explicate contexts make no creative contributions of their own to the whole.

Nevertheless, I must admit that this notion is still open to criticism, because it leads to an infinite regress, in which we attribute a predetermined hierarchical order of levels to an infinite ground of reality that is actually unknown. I now think that it would be better not to make such an assumption but, rather, with regard to any part of the process of development, simply to admit that this context is limited, though in unknown ways. In this sense it would be consistent, for example, to explain the idea that the "vertical dimension" is in some sense finite, but of unknown extent.

We can now consider the question of time and "the timeless"; as explained in TIP, the notion of an implicate order underlying time is suggested by modern quantum mechanical field theory, in several ways. In particular, the quantum mechanical behavior of the gravitational potential implies that neither the distinction between past and future nor that between cause and effect can be maintained unambiguously at distances as short as 10^{-33}. Of course, in Whitehead's point of view, it is admitted that the structure of an actual occasion does not involve time (and that indeed time and space arise only as an order of actual occasions). So presumably there would be no objection to introducing "timelessness" at such short distances, since these may represent the shortest possible actual occasions. However, Whitehead considers longer actual occasions, out of the succession of which entities such as electrons, molecules, etc., are constituted. He also implies still longer actual occasions of still greater duration (as considered in terms of ordinary time) that would be associated with consciousness. Such timeless moments, in principle capable of extending to indefinitely longer durations, were shown in my paper to be able algebraically to contain relationships which effectively are the enfoldment of orders, arrangements, connections, and organizations of elements of process in explicate time and space. For example, it was pointed out that such space-time structure in relativistic geometry has such a "timeless" implicate counterpart. Because of this, it is possible to establish a *relationship* between time and the timeless. Such a relation constitutes a new context going beyond both Whitehead's notion of time as the ordering of actual occasions and the notion, proposed by me in TIP, that time is an unfoldment of the timeless order. In this way, we may hope to do justice not only to the intuitions behind Whitehead's point of view, but also to take into account some of the basic implications of modern quantum mechanical field theory and some of the experiences that human beings have had, throughout the ages, of a "timeless moment."

In this new approach, one no longer implies that the ordinary level of experience has no fundamental kind of significance, nor does one imply that the timeless or the eternal is the only basic reality. Rather, what is crucial is the *relationship* between the two. (In religious terms, this would be the relationship between what has been called the "secular" and what has been called the sacred.") The quantum theory as seen through the implicate order has given an important clue here, in that such a relationship is possible because explicate structures are seen to have "implicate counterparts."

In establishing such a relationship, it is clear that eternity or the timeless should not be considered as absolute. Rather, one may think in terms of what may be called "relative eternity." For example, a moment may have the quality of eternity and yet not cover the whole of reality in full detail. For example, it has been said that Mozart was able to perceive the whole of a composition in such a moment, which was then unfolded in time in all its detail. The proposal is that a similar relationship between time and the timeless may be universal and that we may see it in many areas of experience. Such a relationship may then be the very essence of what is to be meant by freedom and creativity.

Finally, I feel that the conference, and especially the consideration of my work made by Cobb and Griffin (among others), has helped to open fruitful lines of further inquiry. I would like to suggest that what has happened through this conference could be considered, in some way, as an example of how the unfoldment of relationship in process can bring about new insights in the development of theories.

13. David Bohm

TIME, THE IMPLICATE ORDER, AND PRE-SPACE

In this paper, I shall try to bring together the notions of time, the implicate order, and pre-space. The paper begins with a brief résumé of how time may be understood in the context of physics. It then goes on to a consideration of the implicate order in two senses, which may be called "horizontal" and "vertical." The horizontal implicate order will be seen to be close to Whitehead's notion of concrescence and transjection.[1] The "vertical" implicate order, however, brings in some quite different ideas, which raise new and deeper issues. Both types of implicate order together lead on directly to the notion of a pre-space, expressed in terms of algebraic relationships, out of which ordinary space-time emerges as a limiting case. Finally, all these ideas are generalized and extended to a point at which some fundamental philosophical questions implied in this whole subject can meaningfully be raised. Some of these questions are discussed briefly in an Appendix.

13.1. Time in Physics

Time is an abstraction from movement, becoming, and process. This view was implicit in Heraclitus and is also in essence the view of Whitehead, of Prigogine, and, indeed, probably of most of those attending this conference.

As examples illustrating the range of the subject, we may take chemical processes (such as those studied by Prigogine), biological processes, neurological processes, and psychological processes on one side and climatological, geological, stellar, galactic, and cosmic processes on the other. However, I would like to begin by focusing on a very simple physical process, the periodic oscillations of the electromagnetic field in a resonant cavity. If we plot these oscillations on a space-time diagram, and if the cavity is at rest, the nodes will appear as shown in figure 13.1. From these nodes, intervals of time (and space) may be measured and plotted on the coordinates, x and t. This is indeed the most fundamental way known to measure such intervals (e.g., cesium clocks and interferometers).

Now, suppose that this cavity is accelerated to a velocity V. Because Maxwell's equations are Lorentz-invariant, it follows that when the cavity has reached the velocity V, the nodes will have changed. They

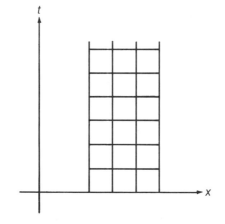

Figure 13.1.

will have undergone a Lorentz transfiguration. When the nodes are plotted in the original space-time frame, the result will look like that shown in figure 13.2. We see from the diagram that the nodes will have suffered a shearing or "squashing" in space-time. However, an observer moving at velocity V and using this also moving cavity as a means of measuring space and time will be treating these "squashed" lines as those of constant time and position. His measurements will produce different values of the space and time coordinates. All the processes in the observer and in his moving environment, being governed by the same laws, will also now proceed in the same relationship to the new coordinates (x', t') as the corresponding processes in the original system do in relation to the old coordinates (x, t). This is the essential content of the special theory of relativity.

It is clear that this relativity in the meaning of time and space coordinates arises because the measure of these coordinates is in fact being abstracted from contexts of processes in two systems that are different in specified ways. Nevertheless, in this theory certain key features are still absolute. The speed of light is an invariant for all the contexts of process under discussion (i.e., for all coordinate frames). The property of being inside the future (or past) light cone is also invariant. From this, it follows that for events inside or on each other's light cones, the order of that which is past and that which is future can never be turned around. This is illustrated in figure 13.3. But for events outside the light cone (e.g., at A), this relative order can be altered (e.g., by means of a Lorentz transformation, A can be changed

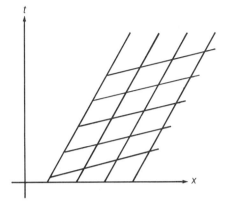

Figure 13.2.

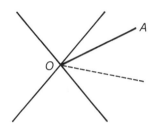

Figure 13.3.

to A'). However, if one assumes that all causal influences propagate inside or on the light cone, then no inconsistencies will arise from this variability in the relative time order of events that are outside each other's light cones.

There is, however, a more important absolute quality to the whole situation: events can in principle be specified in terms of dimensionless points in space-time, and these points are the same in all Lorentz frames. Only their coordinates differ in different frames, being (x, t) in the original system and (x', t') in the new one. This is a key absolute feature of relativity theory, to which I shall return later, when I discuss the quantum theory.

But time has many other meanings in physics. To see what some of these are, let us first consider Leibniz's statement that space is the order of coexistence, whereas time is the order of successive existence. In a nonrelativistic context, coexistence means that one element can act on another that exists at the same time. As we have seen, however, in relativity theory there is no unique meaning to simultaneity, because this depends on the speed of the observer, while contact between events at different positions that would be at the same time in any specified frame is not possible. Strictly speaking, such contact may take place only for events within the same dimensionless space-time point. If we were allowed to consider a pair of events of finite duration, much greater than the time needed for light to connect them, then something close to the ordinary commonsense notion of their coexistence could still be maintained. In view of the requirements of the theory of relativity, one can ask whether this commonsense notion is anything more than the result of imprecise perception and sloppy thinking. Or is it possible, perhaps, in some new way as yet not developed for events of finite duration (and extension) actually to play a fundamental part in making sense of our notions of physical time? I shall return to this question later.

It is a well-known fact that the flux of actual process is irreversible, which seems to indicate a natural order or direction for time. In physics, the basic irreversible feature is that the later time has a higher overall entropy. Increase of entropy has been described as loss of information. Here we must give information an objective meaning, or else we will be implying that irreversibility is somehow merely subjective and ultimately attributable to us.

Gregory Bateson has given a useful definition: information is a difference that makes a difference. Clearly the world is full of objective processes in which there are differences that make a difference, so the basic concept of information is now of objective significance. But are there any actual differences that make no difference? A little reflection will show that our ability to abstract a limited context for study out of a universe of immense size and probably limitless depths of inner structure arises in a very simple way. Its ground is just that the differences in the essentially infinite context that has been left out make no significant difference in the context that has been selected for investigation. This is not just something imposed by us. Rather, it is an objective and indeed invariant feature of the organization of the world that it has all these relatively independent and autonomous contexts, some of which we can discover by suitable inquiry. And as all of this is based on differences that make no difference, we can say that it is the objective loss of information in a given context that allows for the establishment and continued objective existence of the context in question.

Let us now apply these considerations to the entropy. Here we have two relevant contexts, the microscopic and the macroscopic. It is well known from statistical mechanics that there is an enormous number of states specified in a given microscopic context that will correspond to essentially the same state in the macroscopic context. Indeed, Boltzmann's definition of the entropy, $S = k\ln P$ (where k is Boltzmann's constant), leads to $P = e^{S/k}$. Here P is proportional to the number of microstates corresponding to a macrostate. A state of high entropy is one in which large microdifferences correspond to little or no macrodifference or, in other words, to a state in which microinformation is "lost" in the macroscopic context. The increase of entropy with time thus corresponds to a certain loss of information of this kind. In this connection, however, one may add that, since loss of information is what is needed for establishing a new context, the increase of entropy with time may be said to allow later moments to constitute new contexts that are capable of having a certain relative independence from earlier moments.

Such considerations have been criticized on the basis of the assumption that the macrolevel is merely a convenient mode of analysis introduced by us, which means that it is subjective, whereas only the ultimate microlevel is objective and independent of how we think about it. To this, I would answer that what is being proposed here is a theory in which there is no ultimate level. There are only relatively autonomous and independent contexts, in which the differences in what is left out make no significant difference in the context in question. So macrocontext and microcontext are both objective, and yet both are abstractions.

Prigogine (who I believe would agree in essence with this view) and his co-workers have carried this sort of consideration much further, in studies of nonequilibrium processes. They discuss the loss of information as "aging." This is a dynamic and ever-evolving process of inner change in the relationships of microscopic and macroscopic levels, in which microinformation is "lost" in the macrocontext while in the same process the macrolevel may become more organized.

Certainly, aging is one of the basic features of the irreversibility of process and of the universal order in which irreversible process takes place. However, there is more to time than aging. Many systems of different age may be in contact at the same time, while many systems of the same age exist at different times.

In addition, there is evidently more to irreversibility than just loss of information. For example, a very important kind of irreversibility is found in memory. In each moment, there may be enfolded a memory of its past, in which, in turn, there may be enfolded a memory of *its* past, and so on. Memory is thus in a kind of nested order of enfoldment (a bit like Chinese boxes). This is clearly a fundamental kind of irreversibility. However, the possibility of relative independence of the present from the distant past arises because differences in highly enfolded memories will, in general, make little difference within the context of the present. This is somewhat reminiscent of how entropy works. Can these two kinds of irreversibility be related?

In this connection, it is important to note that the full development of the concept of time requires the following elements, to which I have already referred.

1. A recurrent (generally periodic) process, which establishes the measure of time; i.e., what is meant by equal time intervals (e.g., the seasons, the waves in a resonant cavity, etc.). This is clearly reversible.

2. An irreversible process, which establishes an irreversible order of time (e.g., as with the increase of entropy).

3. A fairly stable memory, which enables time intervals in the recurrent (periodic) process to be registered. Such stability of memory evidently requires the irreversibility of time, or else the nested succession of moments could not be maintained, as registered in whatever structure may be the basis of retention of information in memory.

These elements must be related in suitable ways. I shall return to this question later, after discussing the implicate order and pre-space.

Let us now go on to consider the distinction of past, present, and future. In general, what we are conscious of as now is already past, even if only by a fraction of a second. It takes time to produce and organize a meaningful show of this content in consciousness. Moreover, the organization of abstract knowledge takes a lot longer than does that of the relatively immediate sensory experience. The conscious content of the moment is therefore of that which is past and gone. The future is not yet. The present *is,* but it cannot be specified in words or in thoughts, without its slipping into the past. When a future moment comes, a similar situation will prevail. Therefore, from the past of the present we may be able to predict, at most, the past of the future. The actual immediate present is always the unknown. All possibilities of prediction evidently depend on the assumption that the movement is sufficiently slow, regular, and unambiguously related to what comes next, that the difference between the time to which our perceptions and knowledge actually refer and the present makes no significant difference. This is evidently true in a wide range of circumstances. But, in at least two key areas, it is not.

First, in human consciousness itself, what actually happens in its ground may be very fast, irregular, and ambiguously related to the content that can be specified. Therefore, the use of the past to determine what a human being *is* and what he will be can be seriously questioned, if this is carried beyond a certain very limited and probably rather superficial domain.

Second, according to modern physics, microprocesses are also very fast, irregular, and ambiguously related to what comes next. Indeed, it is not in general possible to relate the specifiable information content unambiguously to succeeding events (this is just the essential meaning of the Heisenberg uncertainty relations, as interpreted by Bohr). Here, too, the relevance of the usual notions of time may be questioned. What seems to be called for is that we recognize that the "point event" of relativity theory cannot in general have an unambiguous meaning.

General relativity leads to this conclusion in an even more forceful way. There is Einstein's well-known hypothetical experiment of weighing a photon in a box. To make what is meant more vivid, imagine a box within which is a whole context of process determining a space-time order and measure appropriate to this context. This box is supported on a spring, which has a certain "quantum interdeterminancy" or ambiguity in its height above the ground. It is then a consequence of the general theory of relativity that there is a minimum ambiguity or "uncertainty" in the relationship between the rates of processes inside the box and those outside, which latter are based on the more solid support of a firm foundation that does not move.

More generally, this relationship depends on the whole context. This constitutes a fundamental breakdown in the notion that the point event can be absolute.

Here I think that Whitehead's suggestion of starting with actual occasions having the possibility of a complex inner structure is relevant. But now we add that the relationships of these actual occasions have to have the kind of ambiguity that is characteristic of the quantum theory. In order to emphasize this feature, I suggest that we use the term *moment* (referring here to our actual experience of the moment "now" as never completely localizable in relationship to other moments). This notion of ambiguous (and overlapping) moments is illustrated in figure 13.4. The extension and duration of these moments is in general determined only in some broader context in which they are embedded. In the particular domain covered by the quantum theory, these features will depend on the quantum-mechanical wave function.

And so we see that they are already an implicate order (in which each moment is subject to a certain lack of precise localizability over a region in which the wave function is appreciable).

As we have seen, however, relativity theory describes space-time as completely analyzable down to dimensionless points related by absolute causal laws. In this theory, the whole of space-time has to be considered as a block in which there is no possibility of giving meaning to a moment "now," except perhaps as a subjective illusion. But as we have also seen, quantum mechanics has gone a long way toward the dissolution of the very basis of this picture. It suggests that there *are* objective but context-dependent moments extended in some sense, as well as ambiguously located in space and time. Clearly these may well be significant for understanding what could be meant by "now" as we experience it (which is also extended and ambiguously located in space and time).

Another closely related question is: what is to be meant by *becoming?* If reality is ultimately only the totality of space-time, which is constituted of point events, then becoming has no meaning; it is, indeed, an illusion. But with extended moments it is possible to enfold the past of a given moment as a nested structure (as happens, for example, in memory) that is present in that actual moment. So the moments are not isolated or merely externally related point events but are, instead, internally related extended structures and processes.

Here we may usefully consider the example of a motion picture. Human consciousness does not ordinarily resolve time differences much shorter than about a tenth of a second. The motion picture camera may take as many as ten pictures during this interval, each slightly different from the other. The perception of all of these pictures at once, "fused" in some moment "now," gives rise to a sense of movement and becoming, in which each experienced moment is felt to be "flowing" without a break into the next. But if the pictures are more than a tenth

Figure 13.4.

of a second apart, one senses a series of "jumps" and "jerks" between disjoint images rather than an unbroken flowing movement of becoming. Clearly this means that the sense of becoming is based on the copresence of a number of slightly different images, in any one (ambiguous) moment of consciousness. This sense of becoming also includes a further sense of a certain range of possible lines of development that are to be expected from earlier sequences of images, along with a feeling that the actual development that has taken place, at least in the immediate past, fits in with what is thus implied to be possible.

One can see that this sense of the actuality of becoming must be based on a kind of implicate order in the activity of the brain and the rest of the nervous system. At any moment a certain optical image is on the retina of the eye. The previous image has, meanwhile, been carried by the nerves some distance into the brain, and the one before that yet further, the last image in the sequence being the one that is sensed as present in the moment. The spread of activity that is carrying these images is clearly a kind of enfoldment into the nervous system.

Somehow, this enfolded activity is sensed as movement or becoming. In this enfoldment different images interpentrate and yet make their distinct contributions to the whole impression. The entire experience is rather close to Polanyi's notion of tacit knowing, i.e., knowing in ways that cannot be explicitly denoted or verbalized (e.g., as in knowing how to ride a bicycle). The key point is that becoming is sensed most directly in this highly implicit, tacit, nonlocalizable, and nonspecifiable way. What I wish to propose is that deeply and inwardly *all* becoming, and not merely that which takes place in our conscious experience, is of this nature.

It follows from the above that becoming is not merely a relationship of the present to a past that is gone. Rather, it is a relationship of enfoldments that actually *are* in the present moment. Becoming is an *actuality*. We may enunciate this as a principle, the *being of becoming*. Becoming is being first because in any given moment it is grounded on what is at that moment. Second, it is "being" because the same general structure *continues* in all succeeding moments. (In our example, we see that at each moment there is a set of frames basically similar to the set that is present in any other moment and different only in the details of the content of the pictures, as these change.) The understanding of this requires that we weave together the two principles of the *being of becoming* and the *becoming of being*. I shall discuss what this means in more detail further on.

13.2. The Implicate Order

In the previous section I have already appealed to the implicate order in several ways, in order to help throw some light on the meaning of time. It has been suggested that a key notion in this context is that of a moment (which is close to what Whitehead calls an "actual occasion," as has been pointed out earlier). In each moment, there is a nested sequence of enfoldments of past moments; in turn we may, in another sense, think of this moment as unfolding from these past moments. The moment in question then unfolds into future moments (which in turn will carry forward the same basic pattern, though with constantly different content). This is illustrated in figure 13.5.

Evidently, this notion of enfolding and unfolding is close to Whitehead's idea of the *concrescence* of actual occasions, along with their *transjection* into a set of *consequences* that tends to pervade what follows. This notion is explanatory of the "horizontal" implicate order, which I have mentioned earlier.

However, I would like to give arguments showing that the implicate order has to be extended to many levels and that there is a movement in another context, in which any given level is unfolding (in principle creatively) from levels that are more comprehensive and more fundamental. This series of levels constitutes the "vertical" implicate order.

First I shall discuss this question in general terms. We have seen that each context selected for study will have a certain relative independence and autonomy, based on the fact that, at least up to a point, the differences in broader contexts that have been left out make no significant difference in the context of interests. Of course, the independence and autonomy are limited. Eventually such differences do appear in the context of interest. At first we may regard them as merely contingencies, and if the differences have a random nature we can even treat them statistically, as if they were a further feature of the context

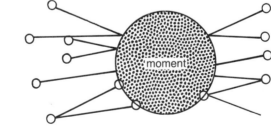

past   future

Figure 13.5.

of interest itself. But eventually these differences will reveal themselves to be more fundamental, in the sense that the basic nature, or indeed even the very existence, of the context under discussion depends on features that cannot be included within this context. This leads ultimately to a new view, in which this context is seen to be embedded in a much wider context, i.e., one that is more comprehensive and, in some sense, deeper. For example, a solid object may at first be disturbed only mechanically by outside forces, but if its temperature is raised it eventually melts or evaporates. This is understood by going to a new context, in which the existence of a solid is seen to depend on a deeper and more universal atomic structure. As we have seen, however, as long as we do not assume that there is an ultimate and fundamental context, the reality of the solid is not denied. The atomic structure and the properties of solids are both real in their contexts, which are, however, limited abstractions from a total reality that cannot be grasped in any specifiable context.

As physics is pursued further, we do indeed find that this atomic structure dissolves into electrons, protons, neutrons, quarks, subquarks, etc., and eventually into dynamically changing forms in an all-pervasive and universal set of fields. When these fields are treated quantum-mechanically, we find that even in what is called a vaccum there are "zero-point" fluctuations, giving "empty space" an energy that is immensely beyond that contained in what is recognized as matter. Other properties, such as polarization, are also attributed to the vacuum. Indeed, the major determinant of the physical content of the theory is just what is called the "vacuum state." What is commonly regarded as matter may then be compared to a set of small waves on the immense "ocean" of the vacuum state.

These "waves" can be shown to form a relatively independent and autonomous context. As happened in the treatment of a solid in terms of its atomic structure, we can say that these waves are realities. Moreover, in a certain sense each wave can be understood as directly related to another such wave. Yet a deeper understanding of such relationships evidently requires that we consider the entire "ocean" in which those waves move and have their existence.

Dirac has used the name *ether* to describe whatever it is that could be meant by terms like *vacuum state* and *zero-point fluctuations*. Perhaps, however, this metaphor (like that of "ocean") brings us too close to the idea of a universal material substance as the ground of all existence. To clarify what is actually meant in this regard by the quantum theory, we should add that the "vacuum state" has to be understood as being an implicate order. Indeed, current mathematical treatments of this state through "graphs" or "Feynman diagrams" may be regarded as particular ways of formulating what is, in effect, a very subtle kind of implicate order (as, indeed, Chew[2] has suggested).

Recently Stapp[3] has further shown that from this overall implicate order one may abstract a particular explicate order, which forms a relatively independent and autonomous context (i.e., not significantly affected by differences in what has been left out of it). This explicate order is based on the part of the overall implicate order corresponding to low-energy protons (i.e., to electromagnetic waves of sufficiently low frequencies). But with the aid of some reflection on this point, we can see that it is in fact just these low-frequency waves that actually permit both the establishment and the measurement of space-time relationships in the context of objects of ordinary size.

As this point is rather important, I want to emphasize it. To repeat, then, the explicate order is a distinguished suborder, based on a relatively independent and autonomous context of low-frequency waves (especially electromagnetic waves). But what is even more important is that the explicate order is no longer regarded as separate, standing over and against an implicate order with which it would be in relationship. Rather, the explicate order *is* a kind of relatively independent "constituent" of the implicate order, with which it fuses and interpenetrates. One may obtain a good image of what is meant here by thinking of this "constituent" as a kind of hologram, which is an implicate "counterpart" of the explicate order. However, within its context the explicate order has a kind of relatively independent reality corresponding to that which we instinctively and unconsciously attribute to "real things" in ordinary experience. But the ground of this reality is the overall implicate order, in which it lies as a "constituent." (I shall go into this in more detail in the discussion of pre-space.)

In a way, the explicate order and its counterpart as a "constituent" of the implicate order are like two views of one object. In the explicate order, all the essential relationships prevailing in the context to which this order applies are displayed in such a way that they stand out, sharply and clearly visible. In a domain in which the implicate context is relatively independent and autonomous, this display is clearly a correct representation of whatever is within this context. But in a broader domain, one has to bring in the dependence of the implicate counterpart on the entire implicate order, and in doing this we are, as it were, seeing the same context from two different views and interpreting these through a single and more comprehensive view.

With all this in mind, let us now go on to consider how time is to be treated in terms of this approach. First, it is necessary to note that in modern quantum-mechanical field theory the "vacuum state" has, properly speaking, no physically meaningful notion of time at all in it. Or, more accurately, in the vacuum state the "state-function" (which represents the whole of space and time) oscillates uniformly at a frequency so high that it is utterly beyond any known physical interpretation. All the physically significant properties of these states are then completely independent of this "zero-point" oscillation. If time has to be abstracted from an ordered sequence of changes in an actual physical process, we would be justified in saying that the vacuum state is, in a certain sense, "timeless" or "beyond time," at least as time is now known, measured, and experienced.

How, then, is actual physical time to be incorporated into the theory? Here it should be noted that in the standard quantum-mechanical treatment, this can be done properly only by bringing in an observer who is outside the quantum system under discussion. (This is usually done through his proxy, in the form of a time-measuring device.) The time of the quantum system has meaning only in relationship to that of the observer. But what if we wish to include the observer (or at least his measuring apparatus) as part of the cosmos? This cannot be done consistently in terms of the usual interpretation of the quantum theory.

However, the work of Stapp indicates, as we have seen earlier, that the entire system of relationships that constitutes the essential meaning of the space-time frame can be contained within the overall implicate order itself. Time then can be understood to be enfolded in a modification of the vacuum state, in which there are a suitable set of low-frequency "waves" constituting matter or light. In this way we may be freed of the need to bring in an external measuring apparatus in order to give a meaning to time.

Let us now develop some of the ideas needed to formulate this notion in more detail. I propose that each moment of time is a projection from the total implicate order. The term *projection* is a particularly happy choice here, not only because its common meaning is suitable for what is needed, but also because its mathematical meaning as a projection operation, P, is just what is required for working out these notions in terms of the quantum theory. A succession of moments is represented by a sequence of projection operators, P_n. The projection of each of these moments is a basically creative act, not completely determined by antecedent moments (though, as we shall see, it is in general related to these moments in a way that is part of a higher order of creativity).

We now let Ψ mathematically represent the state of the totality, which enfolds all space and time in the depths of the implicate order going beyond both of these. We then consider an operation T that transforms Ψ so its elements (or at least some of them) will have a local meaning in an explicate order (i.e., they will be external to each other in this order). The explicate form of the moment is then represented by

$$\varphi_n = P_n T_n \Psi$$

where P_n satisfies the usual algebraic relationship required for a projection operator

$$P_n^2 = P_n.$$

Let us then define another operation that "undoes" the operation T. We call it T^{-1} and by definition

$$T^{-1}T = 1.$$

We then consider

$$\Psi_n = T^{-1}\varphi_n.$$

This operation may be said to "reinject" the explicate form of the nth moment back into the implicate order. This gives Ψ_n, which is just an element in what we have been calling the implicate counterpart of the explicate order. Note that

$$P_n T\Psi_n = \varphi_n.$$

So Ψ_n has the same projection into the nth moment as has the timeless totality, Ψ. However, Ψ_n will be a vanishingly small "constituent" within this totality. It may be thought of as an enfoldment that is something like a hologram of φ_n, which interpenetrates the entire totality while yet being almost infinitely "thin" and "tenuous" in relationship to this totality.

As has been indicated earlier, however, each moment must contain further projections of earlier moments, which constitute a kind of nested sequence of enfolded images of its past. These may take the form of memories. More generally, however, they may be the enfolded "reverberations" of earlier moments within the moment in question (e.g., as in the case discussed previously of cinematic images enfolding into the brain and giving rise to a sense of flowing movement and becoming). Such projection is still to be thought of primarily as a kind of creativity, but here we are discussing *the creation of a moment that is related to its past in a definite way.*

To describe this structure mathematically in more detail, let us suppose that the implicate counterpart, Ψ_n, of the nth moment contains a projection of the $(n-1)$th counterpart, which in turn contains a projection of the $(n-2)$th counterpart, and so on. This is to say, the projection operator, P_n, as it operates on Ψ, will necessarily operate on each "constituent" of Ψ, including Ψ_{n-1}, Ψ_{n-2}, etc. In general, we may expect that the projection of the $(n-1)$th moment as it appears in the nth moment will have "lost" a great deal of original content; when we consider the projection of a moment in the distant past of the nth moment, the result will be extremely "tenuous" even in relationship to that projection which constitutes the essence of the nth moment (as is characteristic, for example, of memories or reverberations of earlier moments in consciousness).

Of course, all these projections into any given moment will have the past of the entire universe as their potential content, which is thus enfolded into the moment in question. In fact, however, moments that are distant in time and space from the one under discussin will generally enfold very weakly (with the result that, as already indicated, each moment can be understood, at least up to a point, in its own relatively independent context).

Let us now consider further how different moments are to be related. Such relationships could, for example, determine general laws that held in broad contexts. These would permit prediction (usually approximate and statistical) of the qualities of later moments in terms of those of earlier moments. But here we recall that knowledge is in the past in any given moment, and from the past in an earlier moment all that we can predict is the past in a later moment. So laws will take the form of generally valid relationships between the nested sets of projections of its past enfolded in one moment and the corresponding set enfolded in another moment. The special creative quality of each moment cannot, however, be predicted in this way.

Of course, all these relationships have to be understood primarily as being between the implicate "counterparts" of the explicate moments. That is to say, we no longer suppose that space-time is primarily an arena and that the laws describe necessary relationships in the development of events as they succeed each other in this arena. Rather, each law *is* a structure that interpenetrates and pervades the totality of the implicate order. To formulate such a law is more like painting a "whole picture" than it is like trying to find a set of dynamical equations for determining how one event follows another. Such dynamical equations will appear only as approximations and limiting cases valid in explicate contexts. Fundamentally, the principles from which the law flows will involve qualities like harmony, order, symmetry, beauty, etc. It has to be kept in mind, however, that breaks in the form of disharmonies, asymmetries, etc., can provide the ground for achieving these qualities anew in a richer way and at a higher level. Perhaps this general feature of the universal order is the reason that the search for symmetries that are later "broken" has been proving to be fruitful in the development of modern quantum-mechanical field theories.

13.3 Pre-Space

It is clearly implied in what has been said earlier that even before we consider the laws of the implicate order we have to provide for the explicate order, on which is based the "ocean" of space and time. As we have seen, Stapp has already indicated a suitable point of departure for beginning actively to carry out such a program. Similar questions have been raised in different ways by others, notably by Wheeler,[4] who has suggested that space-time should be derived as a limiting form of a new notion that he calls *pre-space*. Wheeler's main reason for suggesting this is that the quantization of Einstein's theory of gravitation (based on general relativity) leads to the conclusion that, for very short times and distances, the entire notion of space-time becomes totally undefined and ambiguous. I have already indicated how different systems of time may be ambiguously related, in the discussion of the hypothetical experiment of weighing a photon in a box. But Wheeler is concerned with an even more fundamental ambiguity in the meanings of the terms that define not only the rates of clocks and the lengths of rulers, but also relationships of "before" and "after." The distances at which this ambiguity is significant are very short, of the order of 10^{-33} cm. Yet, when considered with regard to questions of basic principles, such an inability to define "before" and "after" (along with the light cone itself) dissolves the entire conceptual basis on which our notions of space and time depend.

Wheeler has appealed to the image of space-time as a kind of very fine "foam" out of which the familiar patterns and forms of continuous space, time, and matter emerge as approximations on the large-scale level. He is thus regarding this foam as a kind of pre-space, from which ordinary space-time emerges as a suitable limiting case. However, because the structure of the foam is given by quantum laws, one should more accurately regard pre-space as a form of the implicate order.

My attitude is that the mathematics of the quantum theory deals *primarily* with the structure of the implicate pre-space and with how an explicate order of space and time emerges from it, rather than with movements of physical entities, such as particles and fields. (This is a kind of extension of what is done in general relativity, which deals primarily with geometry and only secondarily with the entities that are described within this geometry.)

The fundamental laws of the current quantum-mechanical field theory can be expressed in terms of mathematical structures called "algebras" (indeed, only three kinds of algebras are needed for this purpose, Bosonic algebras, Clifford algebras, and Fermionic algebras). These algebras contain relationships such as addition and multiplication, which enables new terms to be derived from combinations of terms that are already given. For example, one may write

$$A = BC + D.$$

The key point for our purpose here is that any term, K, in the algebra can be transformed into another term, K', by means of a transformation,

$$K' = TKT^{-1},$$

where T itself is part of the algebra. What is particularly significant, then, is that in general this transformation enfolds each term into an implicate order. (That is, K' is a kind of "hologram" of K.) Nevertheless, the relationships between the transformed terms are the same as those between the original terms; or, in the example given above,

$$A' = B'C' + D'.$$

In other words, algebraic relationships are invariant to transformation into an implicate order.

The next step is to connect these algebraic relationships to geometric properties. To go into all the mathematical details involved here would take us too far ahead. However, a more detailed paper is available for those who are interested.[5] In this paper, it is shown that the Clifford algebra (in principle combined with a Bosonic algebra) is able to faithfully portray essentially all the basic geometric properties of relativistic space-time. These include, for example, direction, length, space and time displacements, structures such as lines and their intersections, light cones, triangles, tetrahedra, etc., along with their locations and orientations. This range is wide enough to show that the basic geometric properties of explicate space-time can be mapped into algebraic relationships.

Even more important, however, is the fact that all these relationships can now be put into the implicate order. Or, to put it differently, we can say that an algebraic relationship, such as $A = BC + D,$ corresponding to some explicate geometric features, has an implicate counterpart, $A' = B'C' + D'$, corresponding to the same feature in the implicate order. Although these implicate counterparts all overlap and interpenetrate in the implicate order, their distinctness and their interrelationships are still preserved in an invariant algebraic structure that is characteristic of the quantum-mechanical domain.

With the aid of such algebraic structures, the properties of the explicate order can clearly be related to those of pre-space. For example, one may at first sight wonder how it will be possible to explain the persistence of location and form of an explicate object, which is constantly being created and annihilated in a very rapid succession of projection operations. The answer is very simple. The permanence and constancy of location and form is already being expressed in current physical treatments in terms of standard types of field theories by the requirement that these characteristics (i.e., location and form) be invariant to an *explicate* time displacement D.

But this is an algebraic operation, and so there is another similar operation, $D' = TDT^{-1}$, which is the implicate order. If the implicate counterpart of a geometric property is invariant under the operation D', its explicate form will be invariant under D. Thus, by a natural generalization of the mathematical methods ordinarily used to establish constancy of location and form in the explicate order, we can instead ground these in enfolded relationships in the pre-space.

When this is done the way is opened to go much further, for the ordinary explicate order will now come out only as a relatively invariant context in a much vaster implicate order, containing new features that go beyond those of current space-time and geometrical structures in a radical way.

An interesting example of these new possibilities is afforded by noting that we can now take for granted neither the relative order of moments nor their relative separation, as measured in terms of a time interval. When one puts conditions in the algebra that define these moments within a specified range of uncertainty or ambiguity, one arrives at an expression that is mathematically similar to that which corresponds to the entropy in its current quantum-mechanical representation. The suggestion is that the very possibility of defining a succession of moments in a definite order and with a certain relative separation requires a continual increase of entropy. The way is opened to explore a new relationship between entropy and the fundamental structure that is at the ground of time and space.

Even without introducing new conceptions of entropy in this way, however, it is possible to show that the current quantum-mechanical treatment of entropy already leads to some further insights into time and becoming, when viewed in terms of the implicate order. To do this, I first quote a certain established result of the current quantum theory: when two states of the universe A and B (represented by state functions, Ψ_A and Ψ_B) differ significantly in entropy (by much more than Boltzmann's constant k), the two state functions will be orthogonal[6] (or almost so). In nontechnical language, this means that even though the two-state functions overlap and interpenetrate in the implicate order, no *direct* connection (or almost none) between them will be possible. Adding that two states of the universe with different times will have different entropies provides for the independent existence of two moments of time, in spite of the fact that fundamentally they are not separated in any way at all (it has been noted previously that even classical theory provides for some such independence, but the quantum theory evidently strengthens this a great deal).

If different moments of time are thus (almost) unconnected in any direct way, how are we to understand a process in which one moment *becomes* the succeeding one? The answer is that for a very small entropy change (of the order of Boltzmann's constant, k) the state functions are not orthogonal, so a *direct* connection of one moment to the next is possible if the moments are close enough in entropy. Through a long series of such direct connections, two states of very different entropy may then be connected, as shown in figure 13.6. Thus, for example, A connects with B, B with C, and so on, until finally we reach O, which has no direct connection with A.

It is clear, however, that the above is a good description of a process in which A connects with B, B connects with C, and so on. In this process, each moment *becomes* the succeeding moment with which it is connected. (The time between moments is determined by Boltzmann's constant, k, which implies, in most actual physical processes, a very short time indeed.) But the key point is that process and becoming can be understood in terms of an implicate order, in which vast stretches of time may, in principle, be enfolded. One is thus able to obtain a succession of distinct and not directly connected moments that are *intrinsic* to the implicate order (rather than, for example, having to bring in an external measuring apparatus to do this).

A   B   C   D   E   F   G   H   I   J   K   L   M   N   O

Figure 13.6.

Further enquiry and study are needed to show in more detail how the order of unfoldment at a given level emerges from a "timeless" ground in which there is no separation. For example, we have already seen how, through its projection operator, P_n, each moment enfolds the past projections, P_{n-1}, P_{n-2}, etc. But, of course, the moment in question has further new content, not only because of the implications of general laws relating P_n in more detail to the content of earlier moments, but also because each moment is, in its own right, a creative act. Moreover, if one looks at this process in an inverse way from the standpoint of the future of that moment, one sees that, in the projection of such future moments, the creative action of the moment in question is "carried forward" into these future moments.

The similarity to Whitehead's description of this process is evident, as it contains an equivalent to concrescence, to creativity, and to superjection. But, as has already been indicated, a key difference is that these relationships are grounded in the deeper, "timeless" implicate order that is common to all these moments (which contain the implicate counterparts of all the processes and relationships to which I have referred above). It is this implicate "timeless" ground that is the basis of the oneness of the entire creative act. In this ground, the projection operator P_n, the earlier ones such as P_{n-1}, and the later ones such as P_{n+1} all interpenetrate, while yet remaining distinct (as represented by their invariant algebraic structures). From the point of view of time, however, one can say that this "creative projection" utilizes content supplied in the concrescence of its past and in the very same act provides a basis for what is carried forward as a superjection into the future of the moment.

13.4. Extension and Generalization of Notions of Time and the Implicate Order

I would like now to extend and generalize this approach to time and the implicate order to a point where one can usefully consider some of the deep philosophical questions to which these notions give rise.

First, it is clear that it would be inconsistent with the approach under discussion in this paper to suppose that the vacuum state and its modifications represent the ultimate and unique ground of all reality. Rather, even these are a limited context having only relative independence and autonomy. There will therefore have to be yet deeper and more comprehensive orders within which this context is embedded. The "vacuum state" will thus not be absolute and self-contained. Rather, it too will now be treated as a kind of explicate order, in relationship to a yet more inward implicate order that goes far beyond all that physics has thus far been able to treat (including relativity, quantum theory, the "big bang", etc.). This possibility is indeed already foreshadowed by the fact that in current quantum-mechanical field theory there is available an infinite set of "nonequivalent" vacuum states[7] (this nonequivalence has an origin similar to that of the orthogonality of states of different entropy discussed in the previous section). It will open the way for a new kind of time, which has primarily to do with the development of the vacuum state itself. What I propose, therefore, is a succession of vacuum states, all enfolded in the deeper implicate order, as ordinary time intervals are enfolded in the vacuum state.

On the other hand, this sort of time will be very slow in relationship to the ordinary physical times that are enfolded in it. For this reason, the vacuum state will be subject to a kind of long-range evolution (corresponding roughly to the ancient Greek notion of "aeons"). On the other hand, in relationship to the deeper implicate order, even the immeasurably fast oscillations of the vacuum state to which I have referred earlier may now have a meaning. A new system of time will have been introduced that is both very fast and very slow compared with ordinary physical times. (A good analogy is to consider a radio wave carrying a television program. The very fast radio wave contains enfolded within it the much slower times that are depicted in the program.)

Relativity theory has already led us to expect many different systems of time and space, as these are abstracted from different contexts of process. Quantum theory has led us further to the notion that one system of time and space may be enfolded in relationship to another and that all our common systems are enfolded in the vacuum state. Now we go further to contemplate much greater systems that enfold even the vacuum state with its oscillation and evolution. In all these relationships, any one system has its "timeless" enfoldment in another (or in others). But each system has to be seen in both aspects, i.e., of time and of a relatively "timeless" enfolded state. In the time aspect is comprehended the *becoming of being,* while in the "timeless" aspect is comprehended the *being of becoming.* Since we do not contemplate any ultimate level, it is implied that at any stage in the development of our knowledge these two aspects have always to be woven together for a proper understanding of the nature of time and space. The implicate order is characterized not by an ultimate level in which time is totally absent, but rather by a vast range of interwoven times that enfold other times in nonlocalizable ways and are in turn thus enfolded in still other times.

It has to be remembered, however, that the vacuum state itself will be associated with very short distances (10^{-33} cm). The smallest particles known are much bigger, perhaps of the order of 10^{-16} cm in size. So there is much room for many further interwoven levels of "time" and "timelessness" between these limits. There is no reason to suppose that physics, in its current approach, can exhaust all levels of this kind. Moreover, there may be biological times of various kinds, neurophysiological times, psychological times (both conscious and unconscious) going on, perhaps to levels of which we have at present no notion at all. Each of these levels of time is a relatively independent and autonomous context. Because there is no "bottom level," any given level is not *completely* grounded in any other specifiable set of levels, nor is its explicate aspect as time grounded completely in any particular implicate order of "timelessness." Each level and each aspect makes a certain contribution of its own to reality as a whole.

As long as we restrict ourselves to some finite structures of this kind, however extended and deep they may be, then there is no question of complete determinism. Each context has a certain ambiguity, which may, in part, be removed by combination with and inclusion within other contexts. Any given explicate context will thus not be a *mere* "shadow play" of a deeper implicate context. Everything is ultimately open, as new contexts may always come in. Because of this openness there is a *possibility* for creativity and this possibility will, of course, include the action of human beings. If we were to remove all ambiguity and uncertainty, however, creativity would no longer be possible.

But what if we were to consider, at least for the sake of discussion, the infinite totality of all these levels and aspects? Would this not perhaps leave no room for creativity? This question will be discussed and, as it were, enfolded in what follows.

Let us approach this question by considering our own moments of time, as we consciously experience them. It is clear from what has been said throughout this paper that the explicate and specifiable context of consciousness is based on deeper implicate orders. Most of these are either unconscious or present only as a vague general background in consciousness. Moreover, in analogy to what Stapp has suggested for modern quantum-mechanical field theory, it is natural to suppose that the explicate space and time that we consciously experience is projected from its enfoldment in these deeper implicate orders.

But, as we have seen earlier, such explicate content actually takes time to be organized (e.g., by the activities of the brain and nervous systems), which means that it always refers to that which is past and gone. And so it is partial and cannot represent the totality. What, then, would be involved in being conscious of more of the totality, in the sense of that which is immediately present and actual? To see what this might mean, we may imagine that consciousness sinks into deeper implicate levels, which have thus far been either unconscious or else present as a dimly perceived and vague background. Here consciousness would be mainly of the implicate order, while the explicate order (with its usual sense of time and geometrically organized space) might be a vague and dimly perceived background (thus reversing the ordinary state of affairs). But as we have seen, all such moments "now" have a basically similar structure and differ only in their detailed content. The deeper that consciousness goes into the implicate order, the more similar these moments will then be and the less significant will be their differences. If it were possible for consciousness somehow to reach a very deep level, for example, that of pre-space or beyond, then all "nows" would not only be similar—they would all be one and essentially the same. One could say that in its inward depths now *is* eternity, while in its outward features each "now" is different from the others. (But *eternity* means the depths of the implicate order, not the whole of the successive moments of time.)

There is a well-known saying that "now" is the intersection of eternity and time. On its inward side, "now" is, as we have seen, ultimately the same as eternity. On its outward side, it participates in movement. To go further than this and to ask what the origin of this movement overall is would seem to take us beyond what can be done with thought. Is eternity just something like the holomovement, with no specifiable order or pattern, or does it perhaps ultimately go beyond such description altogether?

What indeed can we say about the infinite totality that would comprehend eternity, time, and their relationship? There are evidently severe difficulties in the attempt to go into this question. Clearly, to try to grasp the infinite in terms of the finite will lead to paradoxes and contradictions. For example, is the totality determinate or not? Evidently, since by hypothesis it already contains everything, there is nothing outside itself that could determine it (i.e., set its limits). This would imply that the totality is completely indeterminate. Yet, because it *is* and because there is nothing else to affect it or to alter it, it cannot be other than it is. And so it is completely necessary and, in this sense, completely determinate. This is almost a tautology. If the totality *is X* (including whatever ambiguity is present in it), then this *X is* the determination of the totality.

Perhaps, however, we can look at this in another way and say that the totality actively and creatively determines itself, in an act that penetrates and pervades all the implicate-explicate levels. This implies that all through the infinite totality there is woven the thread of movement and the thread of being, in the manner that I have already suggested for limited implicate-explicate contexts. In any such context, something new constantly appears: the manifestation of creativity, as previously described in connection with the comparison between the implicate order and Whitehead's philosophy. The new quality can be seen to be grounded in a yet more comprehensive and deeper context. But this grounding is not complete, without also considering the context in question. It is only the whole movement (holomovement), which includes all contexts and their relationship, that is creatively self-determined.

Evidently, the attempt to describe this whole has brought us once again to what one may feel to be a paradox. What is needed, perhaps, is another approach that does not lead us into this paradoxical area. Each level, whether explicate or implicate, has been regarded thus far as a relatively independent and autonomous context *that may be studied in its own right*. But now let us regard all these levels not for their own sake but, rather, for what may be reflected in them. Here we note that reflection is deeply a very active process. Consider, for example, a mirror that reflects light through very fast motions of its constituent change and current distributions. But for us to understand such a reflection properly, we must give primary emphasis not to the context of *the mirrors as autonomous objects* but, rather, to that which can be "seen" reflected in them. So now we no longer take the relatively autonomous contexts of various explicate and implicate orders as the main point of interest. The suggestion is instead that what may be perceived "reflected" in these levels is *just the infinite totality*. Each kind of order thus acts like a mirror, but the two kinds together point more accurately and more deeply into the infinite totality than either alone could do. From this perception consciousness may turn toward the infinite inward depths that are its ultimate ground and, in some way, come into actual contact with these depths.

It may be that only in such an act will the meaning of totality reveal itself. Words may then indicate something of what this meaning is, but only to those who are able to participate in this act. For after all, the word *say* has the same root as the word *see*, indeed, its basic meaning is "to make someone see." There is no way by means of abstract thought alone to "make someone see" the infinite totality (though some thought may well contribute to this). As I have said, a further step is needed in which we cease to look at the content of consciousness, implicate and explicate, for its own sake and regard it as something in which the infinite can be reflected.

Here it is necessary to keep in mind that such questions cannot be approched properly as long as the *questioner* feels himself to be different from a "reflected totality" that he is merely "talking about." Clearly, to have a proper approach requires the cessation of the ordinary analytical divisions between observer and observed, time and the timeless, etc. This will, however, carry us beyond the domain of what is commonly regarded as metaphysical discourse.

A key question, then, is how such an act of perception of unbroken wholeness of the totality will affect the totality (including the human being who participates in this act). Clearly, this question is behind the issues of freedom and the significance of the individual, which, I am sure, deeply concern everyone here. I hope that the consideration of these questions will lead to a fruitful discussion.

Appendix: On Freedom and the Value of the Individual

In order to help lay a basis for the discussion that has been called for at the end of the last section, I shall in this Appendix go in a preliminary way into the background of some of the questions that were raised.

First, let me begin with freedom. Freedom has been commonly identified (especially in the West) with *free will* or with the closely associated notion of *freedom of choice.* In these terms, the basic question is: Is will actually free, or are our actions determined by something else (such as our hereditary constitution, our conditioning, our culture, our dependence on the opinions of other people, etc.)? Alternatively, can we or can we not choose freely among whatever courses of action may be possible?

Such a way of putting the question presupposes that the mind is always able to know what are the various alternative possibilities and which of these is the best. Evidently, however, if one does not have correct knowledge of the consequences of one's actions, freedom of will and choice have little or no meaning. It must be admitted that, in most of human life, lack of knowledge of what will actually flow out of one's choice prevails.

In order to deal with this, we try constantly to improve our knowledge. But as we have seen, reality is infinite both in its depth and in its extension. Although relatively independent contexts do exist, such independence is always limited. Very often the question of what these limits are is obscured not only by ignorance but also by the sheer complexity of the entangled web of interrelationships on which the consequences of our decisions may depend. It often does not seem to be at all likely that we will be able by increasing our knowledge alone even to keep up with this ever-expanding and agglomerating mass of interdependent processes, especially in the field of human relations (consider, for example, what is happening today in politics and in ecnomics).

There has been an enormous expansion of scientific knowledge, especially in the last century, along with a flood of many other kinds of knowledge (this has been called the "information explosion"). But has all this knowledge contributed to our general freedom in any significant way? Or has it not, in many ways, led to yet further unresolvable entanglements in the problem of trying to establish orderly and harmonious human relations? One can mention here the obvious example of how knowledge of nuclear physics has brought us to the need to make all sorts of decisions in situations in which there is little or no reliable information about how human beings will behave (or, even about all the significant physical consequences of our nuclear devices). Without such knowledge no sensible choice is possible. But even if we consider medical knowledge, which is on the whole beneficial (except for the growing number of cases of "iatrogenic" disease), this too is leading to situations in which we have to decide many questions without an adequate basis for making such decisions (e.g., shall terminally ill people be kept alive against their expressed wishes?). In such contexts what do free will and free choice actually mean? This is one of the questions that I hope can be discussed.

Thus far, we have considered lack of knowledge of what may be called the "external world" as a serious limitation on *meaningful* freedom. But there is a much more serious limitation, a lack of what may be called "self-knowledge." Schopenhauer has, at least implicitly, already called attention to this area when he said that though we may perhaps be free to choose as we will, we are not free to will *the content of the will.* Evidently, this content is a key factor determining what sort of person one actually *is,* and yet it appears somehow to be "given." The significance of this question becomes especially clear when we note that people so often seem to be unable actually to do the good things that they have (apparently freely) resolved to do. No person can be said to be free who is for reasons of internal confusion unable consistently to carry out his or her chosen aims and purposes, for evidently such a person is driven by inner compulsions of which he or she is unaware. This inner lack of freedom is far more serious than a lack arising from external constraints or a lack of adequate knowledge of external circumstances.

The problem has often been approached with the aid of moralistic injunctions, telling people to "pull themselves together" and to choose, once and for all, what is right and good. But to those who are unaware of what is actually determining the content of their wills (which includes, in many cases, a content that divides and weakens the will), such advice has little meaning. Moreover, the content is ultimately based on overall self-world views that the individuals have usually not chosen for themselves. These include general notions not only of the sort of world we live in, but also of models of what constitutes a normal right-thinking good human being, of how such human being are to be related, and of what are there duties and obligations, etc. This all-pervasive web of shared thoughts and feelings, propagated not only explicitly but also by tacit and subtle clues picked up since the time of one's birth, operates in most people as an almost overwhelmingly powerful limit on freedom, of which they are essentially ignorant. Indeed, this web not only determines what will generally be thought or felt to be the right choice. Much more important is that it determines what is regarded as the correct range of alternative possibilities in any actual situation. If something is not considered a real possibility, there is no chance at all that it will appear among one's choices. Is there any meaning to freedom of will when the content of this will is thus determined by false knowledge of what is possible, false knowledge that we do not even know we possess (or, more accurately, that possesses us)?

The problem is seen to be even sharper if one considers that the question of choosing what is right and good so often arises in circumstances in which one's desire is in some other direction. Indeed, if something is clearly seen to be right and good, and if one has no desire to do otherwise, it hardly seems that any particular act of choice is ever needed. Will one then not spontaneously have the urge to act according to what one has perceived to be right and good? But often, as has been mentioned above, one finds that one has an irresistible desire in some other direction. One has by no means chosen this desire. Rather, it also arises from the totality of remembrances, reactions, and "knowledge" accumulated over the past, which responds to present "needs" as overwhelmingly urgent in ways of which one is not aware.

The attempt of will to struggle against such desire has no meaning, for this sort of desire contains in it a movement of self-deception, along with a further movement aiming to conceal this self-deception and to conceal the fact that concealment is taking place. Thus, one will often accept as true any false thought that makes one feel better (or more secure) or that makes one believe that the object of one's desire can be realized. This is, for example, the basis of the activity of the confidence trickster, who paints a false picture of satisfying greed that the victim cannot resist accepting as true. As long as one is ignorant of how this sort of self-deceptive desire operates, what can it mean to talk of freedom of any kind?

It appears, then, that the principal barrier to freedom is ignorance, mainly of "oneself" and secondarily of the "external world." This ignorance is also the main barrier to true *individuality*. For any human being who is governed by opinions and models unconsciously picked up from the society is not really an individual. Rather, as has been made very clear, especially by Krishnamurti, such a person is a *particular manifestation* of the collective consciousness of humankind. He or she may have special peculiarities, but these too are drawn from the collective pool of thoughts and feelings (here we may usefully consider the word *idiosyncracy,* which, in its Greek root, means "private mixture"). A genuine individual could only be one who was actually free from ignorance of his or her attachment to the collective consciousness. Individuality and true freedom go together and ignorance (or lack of awareness) is the principal enemy of both.

It is important to note that the main kind of ignorance that destroys freedom and prevents true individuality is ignorance of the *activity of the past.* As has been brought out earlier, although the past is gone, it nevertheless continues to exist and to be active in the present, as a nested structure of enfoldments, going into even the distant past, which are carried along (with modifications) from one moment to the next. (I have treated these as projections of various kinds). Here I have been indicating how this activity of the past can interfere with freedom and individuality as long as one is not aware of this past.

The past is also absolutely necessary in its proper area (as, for example, it contains essential knowledge of all sorts), but when the past operates outside awareness it gets caught up in absurdities of every kind and becomes something like the sorcerer's apprentice, which just keeps on functioning mechanically and unintelligently, to bring about destructive consequences one does not really want.

This brings us to the question: can the past contain adequate knowledge of its *actual activity in the present?* As I have already pointed out, the content of knowledge (which is necessarily of the past) cannot catch up with the immediate and actual present, which is always the unknown. Since the activity of the past is actually taking place in the present, this too is inherently unknown. Thus, knowledge cannot "know" what it is actually doing *right now.* As an example, one may "know" from hearsay or from general conclusions drawn from particular experiences that people of a certain race are inferior. When this "knowledge" responds to a particular member of this race, one's immediate perception is shaped, colored, and twisted so as to present that person as inferior. One is not aware of just how all this is actually taking place, as it happens very rapidly and in very subtle ways. Moreover, the whole process is accompanied by a great deal of distortion and self-deception. For example, as one perceives the "inferiority" of the other person, thus implying one's own "superiority," one experiences a short, sharp burst of intense pleasure. To sustain the pleasure the mind continues with further false thoughts along this line while concealing from itself the fact that it is doing do. Clearly, because the response from accumulated knowledge always lags behind actuality and because, in cases such as this, the mind is caught up in feeling the need to distort, it is not possible through such knowledge alone thoroughly to free the mind of such prejudices.

How, then, is it possible for there to be the self-awareness that is required for true freedom? Along the lines of what has been said in this paper, I propose that self-awareness requires that consciousness sink into its implicate (and now mainly unconcious) order. It may then be possible to be directly aware, in the present, of the actual activity of past knowledge, and especially of that knowledge which is not only false but which also reacts in such a way as to resist exposure of its falsity. Then the mind may be free of its bondage to the active confusion that is enfolded in its past. Without freedom of this kind, there is little meaning even in raising the question as to whether human beings are free, in the deeper sense of being capable of a creative act that is not determined mechanically by unknown conditions in the untraceably complex interconnections and unplumbable depths of the overall reality in which we are embedded.

This leads us to a more fundamental question of what the relationship is between the truly free human being and the totality that goes beyond the explicate order, beyond the implicate order, and beyond time and space. To pose this question is of course itself subject to questioning, since one may ask whether one has properly laid the ground for doing so by giving serious and sustained attention to the actual activity of past knowledge in one's own mind. Is one sufficiently free of this past meaningfully to inquire into the infinite totality? Any real inquiry of this kind must have such attention in it from the very beginning, or else one's mind may be so bound by preconceptions and desires enfolded in one's past that true inquiry at this depth is not actually possible.

Nevertheless, there may be a possibility of usefully engaging, at least to some extent, in a meaningful discussion of this question, even though one inevitably begins from the common state of collective consciousness, which is not properly aware of the actual activity in the present of that false and self-deceptive "knowledge" which is part of its base.

In doing this, the first question to be discussed is: just what is it in our past that binds us, misleads us, deceives us, and thus prevents true freedom? I suggest that what limits us is the attempt to identify oneself with a certain part of one's past that is regarded as essential to what one *is*. As we have seen, we *are* primarily the present, which is the unknown. The past is, in its actuality, merely a part of this present that is also active in the present, and therefore it too is the unknown. All that we know is the *content* (i.e., the meaning) of the past that is gone (and this resembles the actual present only for those contexts in which changes are slow and regular enough so that the difference between past and present makes no significant difference).

If we *are* the unknown, which is the present, then time can be seen in its proper meaning only in the context of that which is beyond time (i.e., the holomovement or eternity). Any attempt to treat the whole meaning of existence in terms of time alone will lead to arbitrary and chaotic limitation of this existence, which then takes on the quality of being rather mechanical. If we are to be creative rather than mechanical, our consciousness has to be primarily in the movement beyond time. Implicitly, this is well known to us. No one will be creative who does not have an intense interest in what one is doing. With such an interest, one can see that one will be at most only dimly conscious of the passage of time. That is to say, though physical time still goes on, consciousness is not organized mainly in the order of psychological time; rather, it acts from the holomovement. On the other hand, if the mind is constantly seeking the goal of finishing its task and reaching its aim (so that it is organized in terms of psychological time), it will lack the real interest needed for true creativity.

Given that a human being may be creative when his or her consciousness arises directly from the "timeless" holomovement, we come to another question: is the creative human being merely an instrument or a projection of the creative action of totality? Or does one act from one's own being independently? I suggest that this is a wrong question, as it presupposes a separation of the human being from the totality, which I have denied at the very outset of this inquiry. A better question is: can we be free to participate in the creativity of the totality at a level appropriate to our true potential?

The need for this question becomes clear if we note that ultimately *everything* is participating creatively in the action of the totality. For matter in its grosser levels, this creative participation consists of continuing to re-create its past forms, with modifications, in a way that is approximately mechanical. (This was implied in the statement that each moment is created with its past as a *projection* containing a further projection of the past of previous moments.) Such creation of a sustained but ever-changing existence of matter at the grosser (mechanical) levels opens the way for the action of higher levels of creativity, such as life and mind.

But once these higher levels are possible, why are they not always fully and harmoniously realized? I have proposed that, at least in part, this is because of ignorance. Such ignorance leads the mind to continue its past, mechanically, through identification rather as if it were a form of matter at a grosser level. The mind is trying in a confused way to realize the kind of creativity appropriate to such grosser levels of matter. In doing so, it is clearly unable to realize the kind of creativity appropriate to its own level.

Ending this state of ignorance may then open a new possibility for the mind to be creative at its own level. When it does this, it is still participating in the universal creativity, but now it is realizing its proper potential.

I suggest that this is the essence of freedom, to realize one's true potential, whatever the source of the potential may be. It is unimportant whether it is grounded in the whole or in some part (e.g., the individual human being). And, indeed, as has been said earlier, the attempt even to raise the question of whether creativity originates in the totality or in the individual presupposes a kind of separateness of the two that we have already denied. So I propose that the question of freedom has to be looked at in a different way.

This new way flows out of giving sustained and serious attention to how unfreedom arises basically from identification with the past, in which the mind commits itself to act as if it were determined mechanically in the ways in which grosser levels of matter are determined. We have to use the past, but to determine *what we are* from it is the mistake. To do this implies that such grosser mechanical existence in time has supreme value and that the main function of the mind is to sustain this sort of existence by continuing the past with modifications. The clear perception that we are the unknown, which is beyond time, allows the mind to give time its proper value, which is limited and not supreme. This is what makes freedom possible, in the sense of realizing our true potential for participating harmoniously in universal creativity, a creativity that also includes the past and future in their proper roles.

What we must inquire into deeply at this conference is, then, the inverse question of what the significance of time and thought is for eternity. This is a very difficult question, and its clarification might well constitute a fundamental step forward in this whole subject.

Notes

1. I use *transjection* as a combination of Whitehead's two terms for this process, *transition* and *superject.*

2. G. F. Chew, "Gentle Quantum Events as the Source of the Explicate Order" (Preprint, Lawrence Berkeley Laboratory 15811); also in *Zygon* 20 (2):1985, 159–64.

3. H. P. Stapp, "Exact Solution of the Infra-red Problem," *Physical Review D* 28 (6):1983, 1386–1418.

4. John Wheeler, *Quantum Theory and Gravitation,* ed. A. R. Marlow (New York: Academic Press, 1980).

5. D. Bohm and B. J. Hiley, "Generalization of the Twistor to Clifford Algebras as a Basis for Geometry," *Revista Brasileira de Fisica,* Volume Especial (July 1984):1–26.

6. G. G. Emsch, *Algebraic Methods in Statistical Mechanics and Quantum Field Theory* (London: Wiley, 1972).

7. Emsch, *Algebraic Methods.*

14. Robert John Russell

A RESPONSE TO DAVID BOHM'S
"TIME, THE IMPLICATE ORDER AND PRE-SPACE."

It is a great pleasure to have been a respondent to David Bohm's paper presented at the 1984 Claremont conference and to have had a chance to write out the response in some detail since then. This is the second time I've interacted in person with Bohm; the first time was in 1983 at the conference sponsored by the Center for Theology and the Natural Sciences (CTNS) in Berkeley, California. On both occasions I have been impressed by his willingness and his ability to respond to questions from widely varying fields of inquiry, to extend and develop his ideas in dialogue with others, and to appreciate and incorporate genuine criticism of his ideas.

Since the CTNS conference, Bohm's research in physics and its implications for philosophy has continued to grow enormously, as evidenced by the paper at hand. This paper is literally packed with new insights and intriguing suggestions, written in Bohm's persuasive and captivating style. My overall response is one of support and encouragement and a sense of being drawn with him through a very subtle and unchartered new territory in physics and philosophy. In this response I will try to mark the points where I think more care should be taken by Bohm, which areas need more immediate attention, and which are most promising in my own opinion. One could also develop a number of theological insights based on Bohm's work, but as this is less related to the central focus of this conference, I will comment on only a few of them. I will focus my response on each section of Bohm's paper and then give a brief conclusion.

14.1. Time in Physics

Bohm starts with the thesis that time is an abstraction from movement, becoming, and process and that it is characteristically irreversible. As an abstraction, it is context-dependent. Bohm supports this thesis with arguments based on relativity, thermodynamics, and quantum physics. Drawing, for example, from special relativity, he argues that the space-time coordinates for the same event will be assigned different values for observers in relative motion and hence that time and space are dependent on the context of observation. Yet the variation of these measurements is just such that, at another level of abstraction, there are features of these events that do not vary. These context-independent features include the speed of light, the order of causal events, and the laws of physics.

This is a valuable insight that relates physics to epistemology, hermeneutics, and other fields. However, it raises many questions which it then leaves unanswered. For example, the theory of relativity has been used by some to support subjective philosophies of nature, whereas others emphasize its objective interpretation, since the theory includes both varying and invariant theoretical elements. The epistemological questions involved are subtle. These questions—including the meaning of the future (is it actual or only potential, determinate or indeterminate?) and the geometrization of time, as Čapek and others have pointed out—deserve more careful attention by Bohm, especially as he continues to develop a realist interpretation of physics.

According to Bohm, temporal irreversibility is normally associated with increasing entropy and hence with loss of information. Again, is such a loss of information objective or subjective? Bohm claims that the concept of information can be made objective if we adopt Gregory Bateson's definition of information as "a difference that makes a difference." Since for Bohm the formation of a context occurs when one abstracts out specific processes from those that make no essential difference to that process, and since the world is organized in an objective sense in terms of such "relatively independent and autonomous contexts," the actual loss of information in the world brings about the formation of context. The passage of time is thus linked to the occurrence of the objective levels of contexts that exist in the world.

This is a striking suggestion. Still, it depends on shifting the question from whether information is objective to whether there exist levels in nature. Such a shift eventually would lead to a decision between a realist and an instrumentalist interpretation of quantum physics. That in turn would involve us in Bell's theorem and the prognosis for a workable theory of nonlocal hidden variables. I do not feel that Bohm's move at this point has helped to settle the issue a great deal.

I agree with Bohm that our conception of time is abstracted from experience and refined by experiment. Human experience is predominated by an irreducible difference between past and future, without which few philosophical, ethical, or theological concepts would be plausible. Hence I welcome arguments for the irreversibility of time that are consistent with physics. Yet Bohm simply accepts the irreversibility of "actual process" as a fact and from this argues to a "natural order or direction for time." But this is actually a very controversial issue. From the standpoint of physics, are irreversible processes at the macrolevel reducible to reversible microprocesses, or is there an irreducible physical basis for time's arrow?

Some look to violations of time symmetry in various weak decays (consistent with Time-reversal Charge-conjugation Parity-invariance) or to the extraordinary "initial conditions" of the cosmos to account for macroscopic temporal irreversibility. More recently, Prigogine and co-workers have argued much more convincingly that temporal irreversibility is intimately associated with nonequilibrium thermodynamics and the production of order out of chaos. This is an outstanding and controversial issue in physics warranting extended technical discussion. Bohm does cite Prigogine in passing, but in my opinion Bohm has done little more than "choose sides." Though I happen to agree with him, I believe his arguments would be strengthened if he acknowledged that it is Prigogine who has given us a solid argument for the irreducible role of irreversible processes in nature. Perhaps on another occasion Bohm will give substantial independent arguments for temporal irreversibility, but I do not find that he has so far.

The problem seems compounded by the overwhelming symmetry of the implicate order. Where would the factors of symmetry-breaking, the "causes" of irreversibility, come from in Bohm's cosmology? Bohm suggests that such asymmetries could be creative and valuable (see below) but, again, is this just a strategic adoption of what in fact is Prigogine's discovery? In general, Bohm's metaphysics, which underlies his discussion of the implicate order, has an inadequate interpretation of contingency, which in turn is a prerequisite for a score of fundamental philosophical and theological issues revolving around free will and determinism. (Some of Bohm's early metaphysics—circa 1956—put even more emphasis on the factors of "chance and necessity" in nature.)

Bohm also links the passage of time to memory, which is described as a "nested order of enfoldment" of a fundamentally irreversible quality. Bohm proposes a parallel between memory and entropy: our sense of the present as distinct from the past depends on "differences in highly enfolded memories [which] will, in general, make little difference within the context of the present" (section 13.1). Here we get a glimpse of another operative thesis in Bohm's world view: the parallel between the structures of physical nature and the cognitive processes of the mind. In this sense Bohm's thought bears a remarkable similarity to Spinoza's. Since both "mind" and "matter" are crucial concepts in Bohm's world view, the relevant issues need more attention.

What of the present moment: is it pointlike or extended and structured, known or unknown? Since it takes time to organize and show experience to consciousness, Bohm argues that "the conscious content of the [present] moment is therefore of that which is past and gone. . . . The present *is,* but it cannot be specified in words or in thoughts. . . . Therefore, from the past of the present we may be able to predict, at most, the past of the future. The actual immediate present is always the unknown" (section 13.1). Quantum theory, with its indeterminacy principle, lends support to this argument, in Bohm's opinion, but this argument depends on whether we adopt a realist interpretation of quantum theory, as Bohm or Heisenberg have argued for, as opposed to the standard instrumentalist position, as Bohr advocated in the Copenhagen interpretation.

Bohm returns to relativity theory, with its view of reality as composed of point events in space-time. He finds such a picture to leave little room for a concept of becoming, and concludes that some new notion of an extended moment is called for in physics. Though I find this conclusion very attractive, his development of it in this paper typifies a problem I have with Bohm. Earlier, Bohm had argued for the sense of the contextuality of time based on relativity. Now he criticizes the assumption of point events found in relativity theory. This would be acceptable if he were evaluating relativity based on his own prior assumptions about time. It is unacceptable if he first uses relativity as a warrant for one assumption, and then concludes that it is inadequate when it later conflicts with another one.

Early in his paper Bohm suggests that Whitehead's notion of an actual occasion, with its complex inner structure, is relevant to physics if one stipulates that such occasions have the kind of internal ambiguity required by quantum physics. Memory, too, with its enfolded past as a present actuality, suggests a different version of reality from relativity theory; in memory, moments are "internally related extended structures and processes." Similarly, our sense of becoming must be related to a form of neurophysiological implicate order. Bohm concludes that all becoming—both physical and mental—is "highly implicit, tacit, non-localizable, and nonspecifiable." Becoming is "a relationship of en-foldments that actually *are* in the present moment. Becoming is an *actuality.*" For Bohm, this issues into a fundamental principle, "the *being of becoming*": "Becoming is being first because in any given moment it is grounded on what is at that moment. Second, it is "being" because the same general structure *continues* in all succeeding moments" (section 13.1).

I think this is a very important development in Bohm's underlying metaphysics. It offers an important suggestion to Whiteheadian phil-osophy: the inclusion of internal ambiguity in the structure of an actual occasion. If this move catalyzes a new connection between process philosophy and both quantum physics and the physics of time irreversibility, it would be a major development in the interaction of philosophy and science.

14.2. The Implicate Order

In the second part of his article Bohm relates his notion of a moment as a "nested sequence of enfoldments of past moments" to Whitehead's idea of the concrescence of actual occasions and their transjection into a set of consequences. Bohm then extends Whitehead's idea, which he sees as corresponding to a "horizontal" implicate order, with the addition of a "vertical" implicate order in which "any given level is unfolding . . . from levels that are more comprehensive and more fundamental" (section 13.2).

The issue here, which dates back at least to the CTNS conference, is complex. At that conference Bohm tended to stress the "vertical" relationship of causality from the implicate to the explicate order. Several members of the conference urged him to give more attention to "horizontal causality" within the explicate order itself. I find a much more plausible balance in Bohm's present metaphysics, in which both vertical and horizontal forms of causality are represented.

But how do the vertical and horizontal orders relate as features of a unified system? In Bohm's current paper, the concept of "context" is developed to interpret the notion of order or level. For example, Bohm interprets the changing description of a solid as it melts in terms of the movement from one context of inquiry to another (mechanics to thermodynamics), thereby suggesting both the limitations of each and the evolution of one into the other in a striking way. Though this is a highly creative and stimulating interpretation, one is left wondering whether the language genuinely reflects an underlying ontology or whether the shifting levels of context are merely descriptive devices.

Bohm goes on to suggest that the vacuum state can be viewed as an implicate order and refers to Feynman diagrams as "particular ways of formulating what is, in effect, a very subtle kind of implicate order." Yet most physicists would treat these diagrams as mathematical prescriptions for simplifying calculational procedures whose only reference is the results of laboratory experiment. Whether one ought to adopt an instrumentalist or a realist interpretation of physics is, in fact, a fundamental issue running throughout Bohm's work, and the problem extends beyond Bohm's work to the whole of the philosophy of science.

Next Bohm cites Henry Stapp's recent results in topological bootstrap theory (TBT). The possible relationship between the TBT program, with its own language and conceptual framework, and Bohm's approach is an intriguing but complex issue, discussed in a paper by Geoffrey Chew last year and again at this (Claremont) conference by Stapp. In this paper Bohm assumes the connection, enabling him to make the following conclusion: "the explicate order is no longer regarded as separate, standing over and against an implicate order with which it would be in relationship. Rather, the explicate order *is* a kind of relatively independent 'constituent' of the implicate order, with which it fuses and interpenetrates" (section 13.2). Bohm's writings have been characterized as monistic, when the implicate order is treated as the only genuine reality, or dualistic, when both implicate and explicate orders are put on an equal footing. If future work sustains his connection with TBT, it could allow a richer metaphysics than either his monistic or dualistic tendencies.

Bohm's writings suggest a variety of theological interpretations as well. If one associates the implicate order with the divine, this statement suggests a form of emanationism. Similarly, when wrestling with the problem of the role of the observer in determining physical time, Bohm contends that, following the work of Stapp, the "entire system of relationships that constitutes the essential meaning of the space-time frame can be contained within the overall implicate order itself." Hence for Bohm "each moment of time is a projection from the total implicate order." This language is again strongly suggestive of the kind of transcendental monism running through much of Bohm's thought. It raises questions, which Bohm works at respondng to, such as how the present can be a genuine locus of creative novelty and yet be a projection of another order. This problem in turn correlates with a complex of theological issues involving predestination, grace, and human free will.

Yet throughout Bohm's comments there are hints of quite different, though still latent, religious themes. For example, the vacuum state of quantum field theory is described as oscillating at a frequency "beyond any known physical interpretation," and Bohm suggests that it is therefore in a sense "timeless" or "beyond time." If Bohm sees the vacuum state as an implicate order, then this argument suggests a deistic theology; if he considers the explicate and implicate orders "like two views of one object," then these sections lean more toward Bohm's tendency to see nature as God, a form of *deus sive nature* that also typifies much of his writing. Clearly, though the theological dimensions to Bohm's work are relatively unexplored, as Bohm's scientific ideas continue to develop and change, their religious implications will require careful tracking.

Another basic part of Bohm's conception of the implicate order is its characteristic harmony, order, symmetry, and beauty. It is these principles which actually produce the structural relationships, such as memory, in each moment or actual occasion. But Bohm now argues that symmetry-breaking can also contribute rich qualities to the explicate order, which raises a question that has concerned me for some time in Bohm's thought: what breaks the symmetry of the whole, determining the irreversibility of time and the one-sidedness of memory? How can the break occur in terms of the symmetric implicate order? This question is packed with implications for many fields of inquiry: for example, aesthetics (is it symmetry or asymmetry which is the source of beauty?), ethics (what promotes the welfare of the whole more, a symmetric or an asymmetric distribution of commodities?), theology (how can evil be real in a world created by a good God?), psychology (what is the basis of self as autonomous agent?), and so on. A labyrinth of issues here will have to be unpacked fully at some point.

14.3. Pre-Space

Wheeler has suggested that macroscopic, continuous space-time be viewed as an approximation because at the Planck length (10^{-33} cm) the effects of quantum gravity make our notions of "before" and "after" meaningless. Bohm suggests that the foamlike structure of space-time at the Planck length be regarded as a form of the implicate order. If it is, quantum formalism is actually a study of the structure of the "implicate pre-space" and the associated problem of how the explicate space-time order emerges from it, not of the movement of particles and fields in space and time.

Bohm supports this idea by an appeal to the algebra of quantum field theory, itself an implicate order, and to the fact that we can map the geometric properties of space-time onto algebraic relationships. From this Bohm can account for persistence of form and location in space-time, since it is grounded in the algebra of the implicate order, and for the relationship between entropy and temporal succession.

If this argument is sustained, it will be an important step for Bohm's research program, which has been attacked on precisely this front. To many, Bohm's previous writing has seemed to lack an adequate account for the regularity and persistence of phenomena and their accessibility to analysis. Just how does the explicate order arise so convincingly out of the implicate order? Bohm's earlier response in terms of the factorization of the wave function was insightful, but by augmenting it with his appropriation of Wheeler's pre-space he may have arrived at a fuller resolution of the problem at hand, as well as another valuable connection to other programs in theoretical physics.

My hesitation, though, is something I have expressed before and can only repeat now. Wheeler's point, it seems to me, rests on the premise that at the Planck length *all* geometric concepts break down, including connectedness, containment, locality, and especially order, and one is therefore forced into a set of assumptions *prior to any kind of geometry.* Consequently, it is hard for me to see how the conceptuality of an "implicate order" can really work, based as it is on geometry (or its equivalent algebraic formulation).

Still, Bohm's discussion of pre-space is filled with insight and tantalizing possibilities for further reflection. Bohm returns to the problem of being and becoming, now in terms of time. According to Bohm, moments of time, though enfolded and interpenetrating in the implicate order, can still be thought of as separated in the explicate order when they differ in entropy. Moreover, they can be smoothly connected by intermediate states approximately equal in entropy. Hence the process of becoming can be understood intrinsically in terms of the implicate order rather than in terms of an external measuring apparatus.

Bohm points out a "key difference" between this description of process and Whitehead's description: the implicate order is the "timeless" ground of all moments that provides for the "oneness of the entire creative act" (section 13.3). By stressing the implicate order as the ground of all occasions, Bohm can account for the unity of experience in a natural way, but again at the risk of writing off the autonomy and concrete actuality of these occasions. Clearly the philosophical issues here will need clarification.

14.4. Extensions and Generalizations

In the final section of his paper Bohm adds another rich insight on being and becoming. The context-dependence of one system of time and space on another, suggested by relativity theory, and the enfolding of one system of time and space into another and finally into the vacuum state, suggested by quantum theory, leads to a further generalization: "any one system has its 'timeless' enfoldment in another" (section 13.4), which means that any system is both temporal and timeless. "In the time aspect is comprehended the *becoming of being,* while in the 'timeless' aspect is comprehended the *being of becoming*." Since there is no ultimate level, the implicate order is like a "vast range of interwoven times that enfold other times in nonlocalizable ways and are in turn thus enfolded in still other times."

I believe that, taken together, the various pieces of this metaphysics, found in the present paper by Bohm, form a remarkable new development in his thinking and represent an important synthesis of many strands in his previous work.

One of Bohm's most admirable qualities, in my opinion, is his continual rejection of reductionism. Here in this section Bohm once again argues against supposing that physics, "in its current approach," would ever exhaust all levels of time. Instead other kinds of time, biological, neurophysiological, psychological, are involved, each as a "relatively independent and autonomous context." With no "bottom level," none of these kinds of time is "*completely* grounded in any other specifiable set of levels." The ambiguity of each context guarantees that "everything is ultimately open." Indeed, for Bohm, creativity is only *possible* if there is "ambiguity and uncertainty."

In this section we also get a sense of the mysticism in much of Bohm's conceptuality. Here he suggests that as consciousness probes the individual "nows" of experience, at very deep levels (such as prespace) these "nows" would be essentially the same. Hence "in its inward depths now *is* eternity, while in its outward features each 'now' is different from the others." He even warns us that a proper approach to these ideas requires "the cessation of the ordinary analytical divisions between observer and observed, time and the timeless, etc." Finally, in the Appendix, Bohm proposes that as our consciousness sinks deeply into the implicate order we will become aware of the activity of past knowledge in the present. This knowledge can free our minds of the "active confusion that is enfolded in its past."

This proposal depends on Bohm's conception of self, a topic which, though not central to this paper, is implicit in much of his writing. Bohm believes that free will depends not only on an adequate knowledge of the genuine alternatives, but even more on knowing how the past is still active in the present moment.

I find Bohm's concept of the past as "active" in the present to be very powerful. It suggests a natural extension into a more attractive metaphysics than that which underlies much of "normal science." It certainly is resonant with the philosophy of organism and with other fields in scientific and philosophic inquiry. Still, I have some considerable disagreement with Bohm's interpretation of self, which to me seems primarily negative.

Bohm believes the past dominates the present, constraining our choice through its "active confusion enfolded in the present." Hence genuine creativity requires one to be freed from the determination of the past. Bohm is admittedly indebted to Krishnamurti at this point, viewing the "principal barrier to freedom [as] *ignorance.*" Though I agree with Bohm that ignorance can be such a barrier, so can knowledge. I would disagree with him that the past is only, even primarily, a source of domination, though it can be. Instead I believe that the past, embedded in the sweeping past of our species, can be a source of vision and an example of the highest ideals of life and that the self, especially in its relation to the self of others, is a venture in discovery and a creation of the divine. Though this is not the appropriate place to develop this response in detail, I have to add that, in my opinion, compassion and forgiveness are the most powerful forces in the *healing* of self and of society and that confession of one's own errors disarms the self-righteousness that knowledge often fosters.

14.5. Conclusion

Bohm has developed several new dimensions to his work that I find extremely interesting. They suggest possible connections to various research programs in physics, philosophy, and theology, and I expect they will continue to be developed by Bohm and others.

Bohm closes with the provocative question: "what is the significance of time and thought for eternity?" This alone is a staggering question. If "eternity" signifies a dimension of the divine reality, I for one believe that as we work openly together on these questions, a better understanding of the gracious divine already in our midst—and of each other in our common struggle—will emerge.

15. Steven M. Rosen

TIME AND HIGHER-ORDER WHOLENESS:
A RESPONSE TO DAVID BOHM

To begin, let me thank David Bohm on two counts: first, for the correspondence we have been engaging in for a year and a half now—I have been enriched by it; second, and more important, for the courageous stand he has taken and the creative leadership he has shown in contributing not merely to advancing the knowledge of his scientific discipline, but to the broader and much more crucial quest for social healing so badly needed at this particular juncture in the history of our species. In my correspondence with David Bohm I have felt a deep harmony of interest and purpose, and the many views we have shared have been a source of gratification to me. We also have discussed some differences in perspective with honesty and openness. My commentary on his paper for our conference is made in the context of our ongoing dialogue.

In his book, *Wholeness and the Implicate Order,* Bohm observes that "to be healthy is to be whole," pointing out that "the word 'health' in English is based on the Anglo-Saxon word 'hale' meaning 'whole' " (1980, 3). There is little doubt that our quest for healing is a search for ways of overcoming fragmentation and achieving wholeness. But this formidable challenge is complicated by the fact that the fragmentation we face is a "fragmentation squared," so to speak. That is to say, the schism that needs to be mended is no simple division of a whole into parts but, rather, a *higher-order* division in which the "parts" themselves are dividedness and wholeness. Although the origin of this currently pathological state of bifurcation long predates the Renaissance, the condition is perhaps most clearly illustrated in the philosophical stance of René Descartes. This position has had a profound and pervasive influence on subsequent Western thought, as Alfred North Whitehead (1978), among others, has amply acknowledged. The "Cartesian split" is in essence a division into that which is divisible—extended in space, thus differentiable, subject to indefinite analysis, precise delineation—and the indivisible—unextended, undifferentiable, thus seamlessly and unanalyzably whole. The domain of the divisble is that of extensive continuity which undergirds the entire program of Western science, mathematics, and technology; it is a domain of sheer juxtaposition, of purely external relations, as Milič Čapek (1961) has incisively demonstrated; in the language of David Bohm, it is the realm of explicate order. On the other hand, the indivisible world is discontinuous, strictly "atomic" or discrete, an impartible unity, a realm of pure *in*ternality of relation, an "enfolded" or "implicate" order—again in Bohm's terms. And by virtue of the *higher-order* nature of the Cartesian split, the differentiated and undifferentiable regimes are themselves differentiated, related in a wholly external manner.

As I see it, the problem of *time* is inextricably linked with that of the higher-order differentiation. I propose that the mystery of process or becoming, from which the familiar idea of time is abstracted, is the mystery of how the undifferentiable becomes differentiated. Time, I suggest, entails a differentiat*ing* of that which is undifferentiable. In more common parlance we may speak of an actualizing, a bringing into extensive continuity, of the purely potential or unextended. The emphasis I have placed on the *ing* suffix in the term *differentiat*ing is to indicate that the undifferentiable and differentiated aspects of time must be understood as related in the most intimate, interpenetrative way. The "paradox" of time is that two distinct features are required, yet these features need to be recognized as one and the same! When they are not, time decomposes into antithetical forms of timelessness (as I will demonstrate). In general terms, I am suggesting that time in its essence is radically non-Cartesian in character and that, therefore, achieving an authentic comprehension of time-as-becoming should entail a perception of higher-order wholeness that could well be therapeutic in its effect. With this in mind, I would like to consider Bohm's approach to the question of time, as well as Whitehead's. I will conclude my presentation by crystalizing the notion of higher-order wholeness in a graphic fashion.

15.1. "Vertical Implicate Order" and "Unknowable Totality": Two Forms of Timelessness

I find myself more or less in agreement with David Griffin's assessment of what he calls in his chapter the "Thomistic and Vedantist moods" of David Bohm. I concur that the former is in evidence in the "vertical implicate order" while the latter shows up in what Bohm refers to as the "infinite totality," "the holomovement," or the "unknown." It appears that both moods end in a denial of the reality of time, though on opposing grounds. Yet both moods *are* needed, because an affirmation of time in the sense of higher-order wholeness I have adumbrated entails their *integration,* as will be seen. It is here that

Bohm assumes a position of leadership in his largely "Thomistic" field. Moreover, in some of Bohm's writings and in our correspondence, I believe there are indications that integration has already begun; I will return to these in the concluding part of my commentary.

If Griffin's interpretation of Thomas as espousing a metaphysics of the finite, i.e., of wholly differentiable form, is correct, the complete *relativization* of the implicate, enfolded, undifferentiable order found in "vertical implicate order" does seem to be Thomistic. That which is implicate ("infinite") in one context is, at a deeper level, in a more comprehensive context, *explicate* (finite).[1] Concomitantly, that which is indeterminable at one level is fully determinable at the next, a notion that seems to render superficial our ideas of freedom and creativity and raise legitimate concerns among Whiteheadians in particular. Similarly with respect to time *per se:* "vertical implicate order" appears to divest time of its genuinely undifferentiable aspect, producing the species of timelessness that Bergson, Čapek, and other process-oriented philosophers have decried—a spatialization of time.

The issue of time and its spatialization has been brought into focus in an especially significant arena for contemporary thought, that of modern physics. Bohm identifies the problem by pointing out that, in the standard approach to quantum mechanics, time has meaning only in relation to an observer who is outside the system. This raises the specter of the subjective—of the concretely, inherently undifferentiable Cartesian psyche or mind—in the very midst of "objective," differentiable physis. And it is this that Bohm, in his "Thomistic mood," wishes to avoid. In order to account for the subjectivistic time associated with the observer, in a strictly *ob*jective manner, this physically undifferentiable implicateness is enfolded in a deeper, *abstract* order by invariant algebraic transformation. It is thereby relativized, rendered explicate, made timeless in the sense of the spatialization of time. So because time cannot be accommodated by physical continuity in the quantum-mechanical system, Bohm introduces a mathematical continuum or "space" whose "dimensionless points" are invariant algebraic operations. In general, it appears that "vertical implicate order" is an indefinitely extended hierarchy in which the appearance of undifferentiability in one context is countered by abstractly invoking a more comprehensive context. Rather than facing the problem of undifferentiability head on—in classical Cartesian language, facing the problem of mind—this problem is put off interminably by an endless sequence of abstractions upon the differentiable or physical. Therefore, "vertical implicate order" does not come to grips with the genuinely, concretely undifferentiable feature of time, the truly atomic aspect of process.

Of course, David Bohm is well aware of this limitation. He *admits* that such mathematical abstractions are only analogies that do not in themselves approach the absolute implicate order, the "infinite totality." And this brings us to Bohm's so-called Vedantist mood. According to Griffin, Bohm sometimes speaks "as if the ultimate reality, the ultimate implicate order, were totally formless." In such a case, "all form and measure would be *maya,* illusion." At times in my personal exchanges with Bohm, I too have gotten the impression of an ultimate denial of form in favor of that which is formless. For example, Bohm has distinguished symbolic knowing from what he believes to be beyond *any* form of thought. Bohm *has* acknowledged that certain forms of symbolizing may usefully call attention to their own limitations and therefore serve as stepping stones, paving the way for transcendence. But in the end, through the acts of inward awareness and deeply reflective attention, which are *distinct* from mere forms of thought, form is entirely left behind; it dissolves in an "intelligent perception of the infinite totality." As I see it, the nonduality thus achieved preserves the higher-order *dualism* of the finite and infinite, the differentiated and undifferentiable, for by granting formlesss totality such priority over form, form does not merely vanish but remains to express itself negatively in the now unsolvable enigma of why there is form at all.

If I am correctly interpreting the paper Bohm has presented for this conference, the same strain of "Vedantist dualism" (along with the enigma it implies) can be detected in the concluding portion (section 13.4) and is particularly evident at the very end of the essay in connection with our central question of time. Here Bohm states: "The clear perception that we are the unknown, which is beyond time, allows the mind to give time its proper value, which is limited and not supreme. . . . What we must inquire into deeply at this conference is, then, the inverse question of what the significance of time and thought is for eternity . . . a very difficult question." This is the "Vedantist" enigma I have referred to. As I finished my reading of the paper with this statement, I was struck by the dualism inherent to its structure and found myself writing in the margin, "But isn't the whole point about wholeness a point about the *non*duality of time and eternity?" Note that whereas the view of time I am proposing requires that time possess nondually, both a differentiated and an undifferentiable aspect, from the "Vedantist" standpoint, "time" is identified with the strictly differentiated and "eternity" with the wholly undifferentiable, thereby giving the second kind of timelessness, the one antithetical to the "Thomistic spatialization of time."

15.2. The "Actual Occasion"

The third corner of the "triangle" for this conference is the philosophy of Alfred North Whitehead. Let me give my current understanding. The Cartesian assumption of *extensive continuity* appears at the deepest level of Whiteheadian metaphysics (see, for example, Whitehead, 1978, 66). As I observed earlier, the relation that obtains between events in an extensive continuum generally is one of mutual externality. The way Milič Čapek has explained it, in a simply continuous space "no matter how minute a[n] . . . interval may be, it must always be an *interval* separating two points, each of which is *external* to the other" (1961, 19). To be sure, in Whitehead's philosophy of process, the fundamental unit is not spatially static but dynamic; of course I am referring to the "actual occasion." Each "actual occasion" is seen as an event that entails the "prehension" of the objective world by the self-determining subjectivity of that moment. In the enfoldment, each "actual occasion" does overcome the separative, simple continuity of dimensionless points abutting in space; with every occasion, extensive space is atomized, the continuous yielding to the discrete. However, the given, momentary occasion itself is *separated* from subsequent occasions in the higher-order, light cone–constrained process continuum that is presupposed. So while every occasion renders the differentiated undifferentiable, the occasions themselves are *differentiated,* external to one another. One way this is expressed in the development of White-headian thought is by the idea that, when the "concrescence" of a given occasion is complete, the subjectivity of this moment has utterly "perished," i.e., the occasion has been totally objectified or "super-jected." In the succeeding occasion, what had just comprised an inward, living subjectivity is now simply external and lifeless, immortalized merely in objectivity, an aspect of the external world to be shaped afresh by a *newborn* subjectivity. The Whiteheadian "rule of thumb" is stated by David Griffin in his contribution to this conference: "subject *then* object" (emphasis added). But to grasp the true working of process, is it really enough for us to view subjectivity merely as prehending other-as-object, i.e., *after* the fact of this other's own living subjectivity? Would not genuine transition demand the *trans*-subjective, core-to-core immediacy of presence of that which is other?

In terms of our focal issue, I venture to say that the Whiteheadian assumption of extensive continuity blocks the attainment of the radically non-Cartesian, higher-order wholeness that I believe is necessary for an authentic comprehension of time as becoming. To reiterate what I see as the fundamental "paradox" confronting us, time is a differentiat*ing* of the undifferentiable. Thus, two distinct aspects are required, the differentiated and the undifferentiable, yet they must be related in a profoundly intimate fashion, not merely viewed as "sides of the same coin" or "end poles of a single continuum," but as interpenetratively one and the same. My understanding is that Whiteheadian metaphysics stops short of this "nondual (monistic) dualism" (see Rosen, 1980, 1983b). The consequence is that, in a subtle way, again we seem to be given an admixture of "Thomism" and "Vedantism." The continuous and the discrete, the well-formed and formless, the differentiated and the undifferentiable are brought into simultaneous view, positioned in close proximity, placed back to back. But they are not convincingly integrated, not fully reconciled, so enigma persists. The mere alignment of opposing forms of timelessness—a completely determined "Thomistic past," an utterly indeterminable "Vedantist future"—does not give us time. Time, I maintain, is the *nondual duality* of the determined and indeterminable.

15.3. The Phenomenology of Temporal Structure

At this point I would like to switch from a purely conceptual approach so as to bring the notion of "monistic dualism" into concrete, graphic focus. My intention is to attain a closer agreement between form and content. Although abstractionism might be a natural form of discourse when the content is Cartesian, I believe the subject matter of the non-Cartesian approach I am advocating additionally demands experiential immediacy for full justice to be done to it. And this should be in keeping with the frequent recommendations of Bohm, who has been a leader in the struggle for recognizing the prime importance of direct intuitive understanding in theoretical science. I will begin with a phenomenological exercise and segue into what might be called "poetic mathematics," an expressive form to which Bohm also has contributed.

My point of departure is a comparison I have employed in several other contexts (e.g., Rosen, 1975, 1977). The divided rectangle of figure 15.1a is a simple illustration of a Cartesian structure. D, the differentiated, and U, the undifferentiable, are represented in abutting squares; juxtaposed, they sit side by side but do not overlap. The connection between them is superficial; they are strictly external to one another. By contrast, consider the relationship inherent in figure 15.1b.

The Necker cube is a well-known figure from Gestalt psychology. Through this depth-creating visual construct, a certain ambiguity of perspective is demonstrated. You may be viewing the form as if it

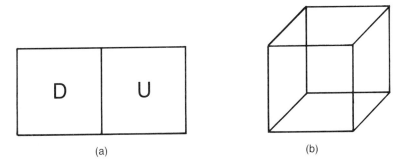

Figure 15.1. The divided rectangle (a) and the Necker cube (b).

were hovering above your line of vision when, suddenly, a spontaneous shift occurs and you are seeing it as if it lay below. This reversal shows, of course, that the figure can be viewed from two distinct perspectives, yet, unlike the parts of the divided rectangle, the perspectives of the Necker cube are not related externally. Each uses the very same configuration of lines to express itself; therefore, the perspectives must be regarded as closely enmeshed. Whereas one square of the divided rectangle easily could be eradicated without affecting the other, no such dissociation of the cube's perspectives is possible. Necker-cube perspectives overlap one another, they do not merely abut.

The proposition I offer is that the divided rectangle be taken as illustrating simple Cartesian structure while the Necker cube—in our *customary* way of viewing it—be interpreted as exemplifying the subtler, higher-order dualism evidenced in Whitehead. Philosopher David Haight (1983) has argued that Whitehead's dipolar monism is a clear improvment over Descartes's dualism, yet still too dualistic to capture "the absolute and actual [as opposed to merely potential] simultaneous presence" of actual entities to each other (Haight's approach is not dissimilar from my own, though there are areas of difference beyond the scope of the present discussion). U and D, the two alternating perspectives of the cube, may be seen to represent the "mental" and "physical" (or "atomic" and "continuous") "poles" of Whiteheadian metaphysics, respectively. True, there is no *spatial* separation of poles, as there is in lower-order Cartesianism, but poles *are* rendered disjoint by assuming a relation of simple succession. Again, the Whiteheadian rule is "subject *then* object," and in this disjunction interpolated between poles a "quantum leap" must be taken from one to the other, leaving us incognizant of what happens in between. In this sense, poles are *completely* polarized, related in a strictly external manner. Interesting and relevant in this regard is Bohm's definition of a mechanistic relation: "entities . . . *are outside of each other* . . . and interact through forces that do not bring about any changes in their essential natures" (Bohm, 1980, 173). Applying this definition to Whiteheadian dipolarity, we may observe that no changes in the essential natures of the continuous and the atomic are brought about by their purely successive interactions. The continuous remains simply continuous or differentiable, the atomic remains simply, undifferentiably that.

Now let us go a step further in our phenomenological exercise with the Necker cube. Rather than allowing our experience of the figure to oscillate from one perspective to the other, we can attempt to view *both perspectives at once* (Rosen, 1985). This can be accomplished by an act of mere abstraction, in which case the cube will simply flatten into an array of connected lines. On the other hand, it is possible to retain the awareness of depth; when this is achieved there is an experience of self-penetration—the form appears to go *through* itself. Such a mode of imaging has a revealing effect on the perception of the cube's faces. In the conventional, perspectively polarized way of viewing the figure, when the quantum jump is made from one pole to the other all the faces of the cube that were seen to lie "inside" presently appear on the "outside," and vice versa. But it is only at "polar extremes" that faces are perceived as *either* inside *or* outside. With the perspectival integration that discloses what lies *between* the poles, each face presents itself as being inside and outside *at the same time*. Therefore, the dualism of inside and out—symbolically, the dualism of the mental and physical—is surmounted in the creation of a "*one*-sided" experiential structure. It must be emphasized that this self-intersecting structure does not merely negate the perspectival distinction between sides, leaving sheer flatness, simple simultaneity (as would an act of mere abstraction). Faces *are* inside, and yet they are *outside* as well. The perspectives of the cube are apprehended both as simultaneous and successive! Thus, the "one-sided" entity produced in the visual exercise with the cube may be said to exemplify the relation of *monistic dualism* required for reconciling the differentiated and the undifferentiable. The result is an insight into the hidden structure of time, the "crack" between simultaneity and succession that is neither alone but both— in fact, nonmechanistically *more* than both, the more being none other than movement, process, becoming itself.

In our final exercise, the proposed solution to the problem of time will be delivered in still more concrete form, the structure being tangibly "materialized," so to speak. We turn to the field of mathematics known as topology, for here a palpable model of one-sidedness can be fashioned. But let me repeat that my intention is not to develop a rigorous mathematical formalism, but to go beyond "Thomistic" rigor to a "*poetic* mathematics." So we will start with topological analogy, then advance—not to axiomatically airtight homology, but in the other direction, to "airy" metaphor, though mathematical nonetheless.

Once more, the point of departure is a comparison I have made elsewhere (Rosen 1975, 1977, 1980, 1981). A cylindrical ring (figure 15.2a) is constructed by cutting out a narrow strip of paper and joining the ends. If one end of such a strip is given half a twist (through an angle 180°) before linking it with the other, the surface of Möbius (figure 15.2b) is formed. In both the cylindrical and Möbius cases, a point on one side of the surface can be matched with a corresponding point on the other. However, in the former case, the pairing is superficial in a mathematical sense. *You* must do the matching, for the fusion of points is not inherent to the topology of the ring. Consequently, points may be regarded as insulated from and external to one another. It is true that, for the Möbius case, by placing your index finger on any point along the strip you will be able to put your thumb on a corresponding point on the opposite side. The paper strip does have two sides like the cylinder. But this holds only for the local cross-section of the strip defined by thumb and forefinger. Taking the full length of the strip into account, we discover that points on opposite sides are intimately connected—they can be thought of as "twisting" or "dissolving" into each other, as being bound up *internally.* Accordingly, topologists define such pairs of points as *single* points, and the two sides of the Möbius strip as but *one* side. Of course, this surface is not one-sided in the merely homogeneous sense of a single side of the cylindrical ring. The local distinction between sides is not simply negated

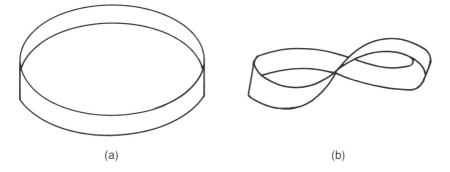

(a)   (b)

Figure 15.2. The cylindrical ring (a) and the surface of Möbius (b).

with expansion to the Möbius surface as a whole; rather, the sides come to *interpenetrate.* Therefore, perhaps a more accurate way to characterize the Möbius relation is to say that it is *both* two-sided and one-sided. I suggest that, in examining the properties of the surface of Möbius, we find a concrete embodiment of the visual experience of perspectival integration performed with the Necker cube, an embodiment of monistic dualism.

But the Möbius strip *is* only an analogy. It *symbolizes* the union of the differentiated and undifferentiable but is itself simply extended in space, subject to precise delineation, wholly differentiable. Now Bohm has written eloquently on the importance of poetry and art in our search for healing (= wholing), i.e., the importance of *metaphor.* In our correspondence, he has pointed out that while mere analogy (simile, prose) does not give us direct access to the implicate, metaphor can:

A metaphor is an *implied comparison.* It is to be distinguished from a *simile,* which is an explicit comparison. In a simile, one says A is similar to B in certain explicit respects, different in others. The elaboration of a simile ultimately gives rise to *prose,* which is an *indefinitely extended simile,* i.e. to talk prose is to say (or imply) that the world is *like* what I say (though also *unlike*). In a metaphor, it is essential that the similarity is left implicated or unstated. I suggest that *the meaning of a metaphor is therefore directly apprehended in an implicate order of the mind.* So in poetry, it is possible to engage the implicate "dance of the mind" (Bohm, 1983).

Bohm's distinction between simile (analogy) and metaphor can be seen as providing a basis for a "poetic mathematics." The key is that, in such a form of expression, "similarity [would be] left implicated." In other words, a "poetic mathematics" would not be predicated on the effort rigorously to delineate, explicate, render invariant, but would accept the radically *non*linear, the implicate or noninvariant, in its very midst. From this standpoint, "vertical implicate order," understood as a hierarchy of invariant algebraic transformations, would have to be regarded as prosaic. But elsewhere Bohm emphasizes another aspect of algebraic representation. He introduces the idea of "nilpotence," i.e., nonzero terms that when multiplied become zero! Bohm observes that "properly nilpotent terms describe movements which ultimately lead to features that vanish [rather than ending in well-delineated products]" (Bohm, 1980, 169). When our aim is invariance, nilpotent terms must be excluded, says Bohm. Conversely, the inclusion of nilpotence should serve the goal of "noninvariance" or "impermanence," i.e., of a metaphorical movement that can directly engage the "dance of mind." In my essay on Bohm's book (Rosen, 1982), I noted that aspects of mathematician Charles Muses' "hypernumber algebra" (1977) seem to serve the same purpose and are closely related to the notion of nilpotence. Moreover, I believe it can be shown that the topological counterpart of the hypernumber epsilon[2] is the *Klein bottle,* the latter being none other than the higher-order version of the Möbius surface, the metaphorization of the Möbius analogy. I propose that the Klein bottle—an "inside-out" structure that *literally* goes through itself—delivers the Necker-cube expression of time-as-authentic-becoming in directly implicate, poetic fashion.

Of course, it is possible to treat such poetic mathematical forms in a prosaic way, to reimpose invariance by the use of abstraction. Indeed, the deeply engrained "Thomistic" habit is to do just that. As Čapek has pointed out, the spatialization of time is an almost irresistible tendency for the post-Renaissance mind. We simply cannot bring ourselves to *accept* impermanence, noninvariance, the flux of process. Instead, we attempt to complete the incomplete by an act of mere abstraction that takes us but another step in an infinite regress. I propose that we can complete incompleteness only by genuinely *accepting* it. Here the "hole" in the Klein bottle (construction of the bottle cannot be completed in three-dimensional space) would not be filled by abstract spatialization, as is the topological convention. Consciousness itself—what Bohm calls "intelligent perception"—would fill this hole. In this way, "Thomism" would be integrated with "Vedantism," the differentiated reconciled with the undifferentiable, the net result being *process.*

So in the final analysis, I believe there is much common ground between Bohm and myself. I would only emphasize my conviction that "Thomistic" and "Vedantist" traditions cannot properly be unified if poetic form is considered a mere stepping stone to formless intelligent perception, to be left behind when the "infinite totality" is realized. In the "poetic mathematics" of which I speak, form and formless intelligence are dually one and the same. It is this monistically dual relation that underlies the paradox of time as process, change, creative advance. Therefore, we may view it as the key to our freedom.

Notes

1. For a discussion of the relativization of the infinite, see Rosen (1983a).
2. The defining characteristics of epsilon are: $\varepsilon^2 = 1$, $\varepsilon \neq \pm 1$.

References

Bohm, David. 1980. *Wholeness and the Implicate Order.* London: Routledge and Kegan Paul.

———. 1983. Personal communication.

Čapek, Milič. 1961. *Philosophical Impact of Contemporary Physics.* New York: Van Nostrand.

Haight, David F. 1983. "Rememberance of Things Present: A Second Copernican Revolution in Consciousness." Paper presented to Association for Transpersonal Psychology, Asilomar, California, June.

Muses, Charles. 1977. "Explorations in Mathematics" *Impact of Science on Society* (UNESCO) 27(1):67–85.

Rosen, Steven M. 1975. "The Unity of Changelessness and Change." *Main Currents in Modern Thought* 31(4):115–20.

———. 1977. "Toward a Representation of the 'Irrepresentable' " In *Future Science,* ed. John W. White and Stanley Krippner, 132–55. New York: Doubleday/Anchor.

———. 1980. "Creative Evolution," *Man/Environment Systems* 10 (5–6):239–50.

———. 1981. "Meta-Modeling as a Strategy for Constructive Change." *Man/Environment Systems* 11 (4):150–60.

———. 1982. "David Bohm's 'Wholeness and the Implicate Order': An Interpretive Essay." *Man/Environment Systems* 12 (1):9–18.

———. 1983a. "The Concept of the Infinite and the Crisis in Modern Physics." *Speculations in Science and Technology* 6 (4):413–25.

———. 1983b. "Paraphysical Reality and the Concept of Dimension." *Journal of Religion and Psychical Research* 6 (2):118–29.

———. 1985. "Can Radical Dualism Accommodate Psi Interaction? A Reply to Professor Beloff." *Journal of Religion and Psychical Research* 8(1):13–25.

Whitehead, Alfred North. 1978. *Process and Reality.* Corrected edition, ed. David Ray Griffin and Donald W. Sherburne. New York: Free Press.

16. Crockett L. Grabbe

AN EXAMPLE OF BOHM'S "IMPLICATE ORDER"

An excellent example of the implicate order can be seen in a phenomenon known as the "plasma wave echo." The experiment proceeds as follows. An external source antenna is placed in a plasma (ionized gas) and a large electric field pulse is imposed. The pulse creates a plasma wave oscillation that rapidly damps away. A short time t later another pulse is applied, creating another wave oscillation that damps away. Immediately after this damping the plasma is back to a normal unperturbed state. There is no measurement technique presently available that could detect any residual disturbance of the plasma. However, phase information on the two pulses is contained in the microscopic velocity distribution of the particles. This information is a property of the whole plasma and is truly an enfolded order. At a time $2t$ this information becomes unfolded as the plasma generates its own pulse from the phase information contained within it. This pulse is the plasma's echo to the first two external pulses.

A review of echoes similar to this is given in Roy W. Gould, "Cyclotron Echo Phenomena," *American Journal of Physics* 37 (1969): 585–97.

17. Ilya Prigogine

IRREVERSIBILITY AND SPACE-TIME STRUCTURE

New ideas about the structure of space and time have appeared at crucial moments in the history of Western science.[1] These moments occurred when new observations coupled with new theoretical concepts shook previously accepted views. Two obvious examples are Newton's synthesis and Einstein's analysis of space-time, which lies at the basis of his relativity theory. Newton's synthesis was preceded by the pioneering work of Galileo, Kepler, and others. Einstein's view on space-time was the crowning outcome of long discussions about the possible effects of the ether wind and of the implications of the celebrated Michelson-Morley experiment. Today we are again in a period of great ferment.

At the start of this century, continuing the tradition of the classical research programs, physicists almost unanimously held that the fundamental laws of the universe were deterministic and reversible. Processes that did not fit this scheme were supposed to be exceptions, merely artifacts due to some apparent complexity, which had itself to be accounted for by invoking our ignorance or lack of control of the variables involved. Now that we are at the end of this century, more and more physicists think that the fundamental laws of nature are irreversible and stochastic, that deterministic and reversible laws are applicable only in limiting situations.

It is interesting to inquire how such a change could occur over a relatively short time span. It is the outcome of unexpected results, obtained in quite different fields of physics and chemistry such as research on elementary particles, cosmology, and the study of self-organization in far-from-equilibrium systems. Who would have believed, fifty years ago, that most and perhaps all elementary particles are unstable? Or that we would speak about the evolution of the universe as a whole? Or that, far from equilibrium, molecules may communicate (to use anthropomorphic language), as witnessed in the chemical clocks to which I shall turn later in more detail?

All these unexpected discoveries also have a drastic effect on our outlook on the relation between "hard" and "soft" sciences. According to the classical view, there was a sharp distinction between simple systems, such as those studied by physics or chemistry, and complex systems, such as those studied in biology and the human sciences. Indeed, one could not imagine greater contrast than that existing be-of a gas or a liquid and the complex processes we discover in the evolution of life or in the history of human societies.

Still, this gap is now being filled. Over the last decade, we have learned that, in nonequilibrium conditions, simple materials, such as a gas or a liquid, or simple chemical reactions can acquire complex behavior. This unexpected complexity leads to the appearance of broken temporal or spatial symmetry, to chaos, and to patterns of increasing diversity.

In addition to nonequilibrium physics, an outstanding role in these new developments is played by the revival of the theory of classical dynamical systems. Here also a dramatic change took place. We know now that the simple motions, whose descriptions still fill most textbooks of classical dynamics, correspond in fact to exceptional situations. Whenever we shift to other situations, such as from two-body problems to three-body problems, the complexity inherent in dynamical systems increases dramatically. I want to show in this essay that the recent progress in nonequilibrium physics coupled with that in dynamical systems leads to fundamental changes in our concepts of space and time that may be of great importance over practically the whole range of basic physics, be it classical mechanics, quantum theory, or relativity. Let us start with some considerations involving nonequilibrium physics.

17.1. Dissipative Systems

The example I shall use corresponds to thermal convection, which is explored today in many laboratories because its importance extends from simple laboratory experiments to large-scale phenomena in atmospheric and planetary physics. Consider a fluid layer heated from below. We start first with the system in thermal equilibrium; the temperature is then uniform, and all points inside the layer have the same properties. We then apply an external constraint (see figure 17.1a) and make the temperature in the lower face slightly higher than the temperature in the upper face. Obviously there is now a vertical gradient of temperature due to the external constraint, but all points that lie on the same horizontal plane are equivalent (figure 17.1b). However, as we continue to increase the difference in temperature, the well-known Bénard instability occurs. We observe, in addition to thermal conduction, convection bringing "warm" molecules to the cold wall. This convection may give rise to rolls. Now spatial homogeneity is broken. Two points in a given plane are no longer equivalent. Moreover, if the motion is clockwise in one roll, it is counterclockwise in the next one (figure 17.1c). We expect, therefore, that there will be a basic stochastic element. When we repeat the experiment, there may appear at a given point a left or a right roll. The existence of these rolls implies long-range correlations, as there are about 10^{21} molecules participating in a single roll. If we continue to heat the roll phase, we obtain new instabilities leading to more complex turbulent types of structures. This is of no importance for us here, as I want simply to emphasize that irreversibility may be at the basis of symmetry breaking.

Today also we know many chemical clocks corresponding to strongly nonlinear autocatalytic systems. The simplest experiment corresponds to experiments in stirred chemical reactors. If the residence time is short enough, the sectors may have the appearance of periodic behavior—the so-called chemical limit cycle. Obviously, the appearance of such a cycle breaks the homogeneity of time. However, this is only one of the possible types of behavior we observe. There may be more complicated temporal behavior expressed by chaotic attractors (see figure 17.2).

Let us reflect on what characterizes dissipative systems such as occur in hydrodynamics or in chemistry. I believe there are two main properties that should be emphasized. One is the existence of "attractors." The simplest example, well known to everybody, is the pendulum with friction. If perturbed from the equilibrium position, which is here the attractor, the pendulum goes back to the equilibrium position. However, the attractors may be more complicated than are isolated points. There may be lines in phase space such as in periodic chemical reactions or even more complicated mathematical objects like fractals. That is why one speaks today of "strange attractors." The second common element in dissipative systems is the dissymmetry in respect to time. All dissipative systems have a preferential direction of time; they progress toward their attractors for t going to $+\infty$ (and not for t going to $-\infty$).

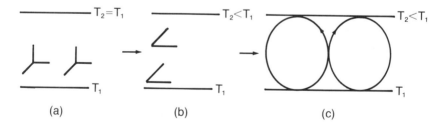

Figure 17.1. Bénard instability: (a) uniform temperature, (b) layer heated from below, (c) layer beyond the Bénard instability.

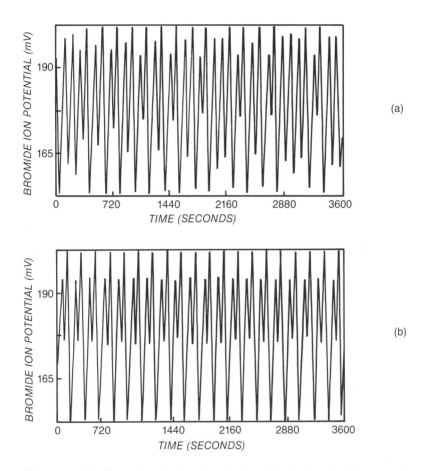

Figure 17.2. Oscillatory chemical reactions in stirred chemical reactor. The example corresponds to the Belousov-Zhabotinskii reaction: (a) chaotic state; (b) complex periodic state.

We could imagine a world in which biological systems belonging to the class of dissipative systems would age while others became younger. In such a world some dissipative systems would tend to equilibrium for $t \rightarrow +\infty$ while others did so for $t \rightarrow -\infty$. But that is not our world, in which, so far as we know on empirical grounds, there is a universal time asymmetry.

Let us now turn to conservative systems.

17.2. Conservative Systems

In classical mechanics conservative systems would be described by Hamilton's equations of motion. From the abstract point of view, such systems are characterized by a phase space and by a measure that is preserved in time. A striking difference between the two is that dynamical systems are never stable in the same sense as dissipative systems. They do not have the property of asymptotic stability. When we give a larger-amplitude oscillation to a frictionless pendulum, it takes up a new frequency and conserves it as long as friction can be neglected. In contrast, when we run, our heartbeat increases but returns to normal after we take a rest. As a result, in comparison with dissipative systems, dynamical systems are basically unstable. There is no way of forgetting perturbations. As a result, the world of conservative dynamical systems is certainly not a world in which delicately balanced processes such as we see in biology would be possible.

Let me illustrate this remark by considering first the classical example of a pendulum. The representation of the trajectories in phase space is well known (see figure 17.3). Point E corresponds to the equilibrium position. Points H_1 and H_2 are in fact identical and correspond to the unstable situation in which the pendulum stays on its head. Point E is an "elliptic" point. If the system is in the neighborhood of E and we perturb it, it will shift its trajectory to a new periodic motion around E. In contrast, points H_1 and H_2 are "hyperbolic" points corresponding to the crossing of a stable and an unstable trajectory called the "separatrices." The separatrices S_1 and S_2 separate the region of vibration

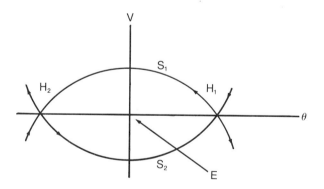

Figure 17.3. Phase space of a pendulum. Here V is the velocity and θ the angle of deflection; S_1 and S_2 are the separatrices; E is an elliptic point; H_1 and H_2 are hyperbolic points.

around equilibrium from the region of rotation. The elliptic point *E* is *orbitally* stable in the sense that the perturbed system remains on a *neighboring* orbit, whereas the hyperbolic points are unstable. In classical examples such as the pendulum there are generally only a few hyperbolic points. On the contrary, there is in most systems that are at present at the center of interest in dynamics a multitude of both elliptic and hyperbolic points. A very simple example I shall use for illustration corresponds to the baker transformation.

In the baker transformation we take a square and flatten it into a rectangle, then we fold half the rectangle over the other half to form a square again. This set of operations is shown in figure 17.4, and may be repeated as many times as one likes. Each time the surface of the square is broken up and redistributed. The square corresponds here to the phase space. The baker transformation transforms each point into a well-defined new point. Although the series of points obtained in this way is "deterministic," the system displays in addition irreducible statistical aspects. Let us take, for instance, a system described by an initial condition such that a region *A* of the square is initially filled in a uniform way with representative points. It may be shown that after a sufficient number of repetitions of the transformation, this cell,

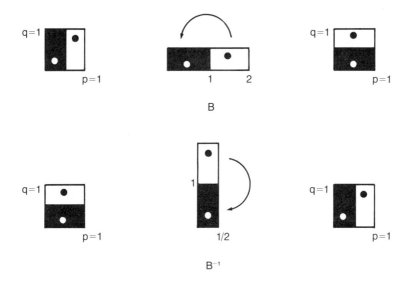

Figure 17.4. Realization of the Baker transformation (*B*) and of its inverse (B⁻¹). The path of the two spots gives an idea of the transformations.

whatever its size and localization, will be broken up into pieces (see figure 17.5). The essential point is that any region, whatever its size, thus always contains different trajectories diverging at each fragmentation. Although the evolution of a point is reversible and deterministic, the description of a region, however small, is basically statistical.

A characteristic feature of the baker transformation is that each point corresponds to the crossing of two orthogonal lines, one vertical, which corresponds to a contracting fiber, the other horizontal, which corresponds to a dilating fiber. Each point therefore corresponds to a hyperbolic point. There is also an abundance of elliptic points; however, they measure zero in the same sense as the measure of *rational numbers* is vanishing. Still, we would expect to find in the baker transformation what is often called "orbital randomness." According to whether we start with an elliptic or with a hyperbolic point, we would observe quite different behavior.

There is also another very important concept involved in unstable dynamical systems. That is the concept of a Lyapounov exponent or a Lyapounov time. In Lyapounov systems the distance δ between two points increases exponentially with time:

$$\delta_r = (\delta_r)_0 \exp^t.$$

τ_l may be called the "Lyapounov time." (Here I consider an *average* Lyapounov time.)

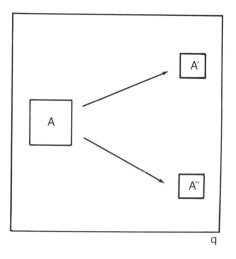

Figure 17.5. Evolution in time of an unstable system. As time goes on, region *A* splits into regions *A'* and *A''*, which in turn will be divided.

Nature presents us with two types of dynamical systems: dissipative systems and conservative systems. What is the relation between the two? This question has intrigued physicists for over one hundred years. It is a difficult question, but I believe that we are now coming closer to the answer.

17.3. The Search for Unification

Thermodynamics provides us with a fundamental insight on the difference between dissipative systems and conservative systems. Indeed, the second law of thermodynamics introduces a basic new quantity, the entropy S, which is fundamentally related to dissipation. Dissipation produces entropy. But what, then, is the meaning of entropy? Over a century ago Boltzmann came up with a most original idea: entropy is basically related to probability; that is

$$S = k \log P.$$

It is because the probability increases that entropy increases.

Let me immediately emphasize that in this perspective the second law would have great practical importance but would be of no fundamental significance. In his excellent book, *The Ambidextrous Universe,* Martin Gardner writes: "Certain events go only one way not because the can't go the other way but because it is extremely unlikely that they go backward."[2] By improving our abilities to measure less and less unlikely events, we could reach a situation in which the second law would play as small a role as we want. This is the point of view that is often taken today, but it is difficult to maintain in the presence of the important constructive role of dissipative systems, which I have emphasized. In fact, we here confront a problem quite similar to the famous quarrel about hidden variables, to which physicists have devoted so much time in recent years. Is probability the outcome of our ignorance, of our averaging over hidden variables? Or is probability genuine, expressing some nonlocality in space-time? In the case of quantum theory the answer is at present quite clear. The introduction of probability in quantum theory results from our use of Planck's constant, h. Interestingly, in the problem of irreversibility we shall also come to the conclusion that the probability is genuine and that the underlying nonlocality is a reflection of the instability of motion of the dynamical systems to which the second law of thermodynamics can be applied.

However, there seems at first to be a basic difficulty in attempting any unification between thermodynamics and dynamics. In dynamical theory we can introduce a distribution function ρ that evolves in time according to the law

$$\rho_t(\omega) = U_t \, \rho_0(\omega),$$

where U_t is a unitary operator (it is often written as $\exp(iLt)$, where L is the so-called Liouville operator). The unitary operator U_t satisfies the group relation

$$U_t U_s = U_{t+s}, \qquad t,s \gtrless 0.$$

As a result of the unitary character of dynamical evolution, entropy, which is expressed as a function of ρ, remains unchanged in the course of dynamical evolution. This is in striking contrast with what happens with probabilistic processes, such as Markov chains. There also we may express evolution in terms of some operator W_t acting on the initial distribution function:

$$\tilde{\rho}_t(\omega) = W_t \, \tilde{\rho}(\omega).$$

However, this new operator is no longer unitary and now satisfies the semigroup condition

$$W_t W_s = W_{t+s}, \qquad t,s \gtrless 0.$$

The problem of unification of dynamical systems and dissipative systems is essentially the problem of elucidating the relation between ρ, corresponding to dynamical evolution, and $\tilde{\rho}$, corresponding to the evolution of probabilistic processes. I cannot go into details here,[3] so let me simply mention that it has been shown that it is possible to go from the dynamical distribution to the probabilistic one in terms of the transformation

$$\rho = \wedge \tilde{\rho},$$

where $\wedge$ is a suitable operator breaking the time symmetry and introducing a nonlocal description in space-time. In other words, in dissipative systems the fundamental laws are no longer the laws of dynamics, as we must include the second law of thermodynamics. As a result, the basic object evolving in dissipative systems is no longer the initial distribution function ρ but a transform of this distribution function, the transformation itself being determined by dynamical laws. Dissipative systems correspond, therefore, to a new level of description.

A historical analogy may come to mind. In the early days of statistical mechanics, the Ehrenfests emphasized the need to introduce a coarse-grained distribution function that would satisfy the second law, in contrast with the fine-grained distribution ρ, which satisfied the laws of dynamics. This idea is the basis of well-known textbooks, such as the outstanding one by Tolman on statistical mechanics.[4] However, the coarse-grained distribution was considered the outcome of ignorance of the nature of the fine-grained distribution and as resulting from some average procedure as applied to the fine-grained density. This cannot be the whole story, as the arbitrariness of the coarse-grained distribution would lead to arbitrariness in the temporal evolution of dissipative systems, which is not borne out by experiment. Moreover, the coarseness of the new distribution must include the breaking of symmetry of the initial equations of motion. Anyway, our method of unification of thermodynamics and dynamics can be viewed as a way of adding precision to the intuitions of the founders of statistical mechanics.

17.4. From Dynamics to Thermodynamics

As expressed in the preceding section, in my view the transition from dynamics to thermodynamics corresponds to the transformation from the dynamical fine-grained distribution ρ to the distribution ρ̃, which exhibits a Markov-chain property. The existence of this transformation can be rigorously proved for an important class of dynamical systems that are highly unstable, such as the baker transformation. It can also be proved, but by perturbation techniques, for a larger class of dynamical systems, such as are described usually in terms of kinetic equations. Let me emphasize the two elements necessary for this transformation.

The first is, as already mentioned, a high degree of instability. It is instability that makes the concept of trajectory unphysical and that leads therefore to the very possibility of applying probability concepts in an objective sense. Systems having a high degree of instability may be called "intrinsically random." But there is a second element. Only for systems in which the symmetry of time is broken can we hope to formulate the second law of thermodynamics. In other words, it is only when there exist states of motion whose velocity inverse is forbidden that the transformation from dynamics to thermodynamics makes sense. This means that the second law can be expressed as a selection principle that selects only those initial conditions which are compatible with the approach to equilibrium in the distant future.

Let me now illustrate the idea of the second law as a selection principle. As already mentioned, at every phase point there are now two manifolds (of lower dimensionality than the entire phase space): one that progressively contracts under dynamical motion for increasing t, and the other expanding with t. As the result of the baker transformation represented in figure 17.6, a vertical line will progressively contract to smaller and smaller vertical lines under successive applications of the baker transformation ("contracting" fibers), whereas a horizontal line will double with each application of the baker transformation ("dilating" fibers).

The contracting and expanding manifolds, when they exist, are *evidently time-asymmetric objects*. The contracting manifold moves, in a sense, as a single unit toward the future. All its points tend toward the *same fate* in the future, but they can be seen to have diverging histories as we look further into the past. Expanding manifolds are just the opposite. Points on an expanding manifold will have diverging future behaviors but progressively converging histories.

It is the existence of such time-asymmetric objects that enables one to construct the symmetry-breaking transformation $\wedge$ by assigning *nonequivalent roles to expanding and contracting manifolds*. In fact, it can be shown that the choice of $\wedge$, which gives rise to entropy-increasing evolution (for $t \geqslant 0$) as the physically realized symmetry-breaking transformation, implies the exclusion of (singular) distribution functions concentrated on contracting manifolds from the set of physically realizable states.

What is physically realizable and what is not is, of course, an empirical question. What our formulation of the second law achieves is to link

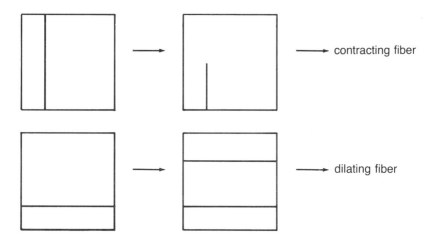

contracting fiber

dilating fiber

Figure 17.6. Baker transformation—contracting and dilating fiber.

the second law and the associated "arrow of time" with a limitation, on the fundamental level, of preparing certain types of initial conditions. In physically interesting models of dynamical systems, the types of initial conditions that are excluded by the symmetry-breaking transformation $\wedge$ are precisely those which one would intuitively expect to be unrealizable.

Many examples can be given. In scattering theory we may have a plane wave giving rise to an outgoing spherical wave.[5] We could have also an incoming spherical wave transformed into a plane wave. Both processes are strictly symmetrical from the point of view of the laws of dynamics. However, only one of these two types of phenomena occurs in nature. It is very gratifying that the selection principle as included in the second law of thermodynamics has a very simple physical meaning: we cannot prescribe a common future for ensembles in unstable dynamical systems.

I shall now describe more precisely how to achieve the construction of the $\wedge$ transformation.

17.5. Internal Time

For the class of unstable systems of which the baker transformation is the simplest example, the construction of $\wedge$ may proceed most directly through the consideration of what Misra and I have called the internal time T.[6] To grasp the intuitive meaning of internal time, think about a drop of ink in a glass of water. The form the drop takes gives us an idea of the interval of time that has elapsed. We may also consider the baker transformation and look how ink will be distributed in the square as a result of successive transformations. This succession is represented in figure 17.7. The shaded regions may be imagined to be filled with ink, the unshaded regions with water. At time zero we have what is called a "generating partition." From this partition we form a series of either horizontal partitions, as we look into the future, or vertical partitions, as we look into the past. These are the basic partitions. An arbitrary distribution of ink in the square can be written formally as a superposition of the basic partitions. To each basic partition we may associate an "internal" time that is simply the number of baker transformations we have to perform to go from the generating partition to the one under consideration. We therefore see that this type of system admits a kind of internal age.

The internal time T is quite different from the usual mechanical time, since it depends on the global topology of the system. We may

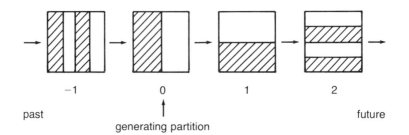

past   future

generating partition

Figure 17.7. Starting with the partition at time 0, we repeatedly apply the Baker transformation. We generate horizontal stripes in this way. Similarly, by applying the Baker transformation in the other direction from time 0, we obtain vertical stripes.

even speak of the "timing of space," thus coming close to ideas recently put forward by geographers, who have introduced the concept of "chronogeography."[7] When we look at the structure of a town or a landscape, we see temporal elements interacting and coexisting. Brasília or Pompeii would correspond to a well-defined internal age, somewhat like one of the basic partitions in the baker transformation. In contrast modern Rome, whose buildings originated in quite different periods, would correspond to an average time exactly as an arbitrary partition may be decomposed into elements corresponding to different internal times.

It is very important to notice that the internal time corresponds now to an operator whose eigenfunctions are the partitions, in the case of the baker transformation, and whose eigenvalues are the times shown on a watch. In this simple example, change in the average internal age $<T>$ keeps track with time on the watch. But for more complex dynamical systems the relation between the average time $<T>$ and watch time becomes more complicated. Obviously, internal time corresponds to a nonlocal description. If we knew exactly the position of a point in the square corresponding to the baker transformation, we would not know the partition to which it belongs. Inversely, if we knew the partition we would not know the position of the trajectories.

The question of what time is has fascinated human beings since the dawn of modern thought. Aristotle associated time with motion, but he added: "There must be also a soul which counts." In a sense the "soul which counts" is replaced here by internal time, which is measured by astronomical time but is not identical with it. A watch has no time in our sense. Everyday it goes back into its own past. It is we who have time, and this time expresses the fact that, like all chemical systems, we belong to the category of highly unstable dynamical systems for which an object such as T can be defined. The prototype of the physical world in classical thought was planetary motion. Now, I believe, the prototype is the kind of highly unstable dynamical system from which internal time and irreversibility may be generated.

17.6. States and Laws

In the preceding section I introduced the transformation function $\wedge$ and demonstrated that it is simply a decreasing function of the internal time $\wedge(T)$. This fact has some very interesting implications. Suppose we expand the fine-grained distribution function ρ in terms of the eigenfunctions χ_n of the internal time T. We obtain formally

$$\rho = \sum_{n=-\infty}^{+\infty} c_n \chi_n.$$

If we now apply the transformation $\wedge$ we obtain, similarly,

$$\tilde{\rho} = \wedge\rho = \sum_{n=-\infty}^{+\infty} c_n \lambda_n \chi_n.$$

The λ_n values are the eigenvalues of operator $\wedge$ corresponding to the eigenfunction χ_n, with the important requirement that λ_n varies from 1 for $n \to -\infty$ to 0 for $n \to +\infty$. Here n is the eigenvalue of the internal time. Both ρ and $\tilde{\rho}$ at a given time are made up in general from contributions from both past ($n < 0$) and future ($n > 0$) in terms of the internal time T. However, while in ρ future and past play a symmetrical role, this is no longer so in $\tilde{\rho}$. In $\tilde{\rho}$ the contribution of the future states is damped as λ_n goes to zero for $n \to \infty$. The present contains the contributions from the past and contributions from the nearby future. This is in contrast with dynamical deterministic systems, where the present implies both the past and the future.

Let us represent λ_n as a function of n (see figure 17.8). Past and present are separated by a kind of transition layer. It may be shown that this transition layer is of the order of the Lyapounov time, τ_L. It is interesting to contrast this representation with the traditional representation of time as a straight line (see figure 17.9). The present in the traditional view corresponds to a single *point* that separates past from future. The present comes, so to speak, out of nowhere and disappears into nowhere. Moreover, being reduced to a point, it is infinitely contiguous to the past and the future. In the representation I propose,

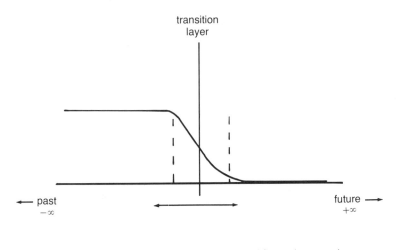

Figure 17.8. Transition between past ($n \longrightarrow -\infty$) and future ($n \longrightarrow +\infty$).

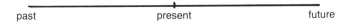

past   present   future

Figure 17.9. Traditional representation of time.

the past is separated from the future by an interval measured by the Lyapounov time: we may speak of the *"duration"* of the present.

It is interesting that many philosophers, especially Bergson[8] and Whitehead,[9] have emphasized the need to attribute to the present some kind of incompressible duration. The view of the second law as a dynamical principle leads precisely to this conclusion.

From the classical point of view, initial conditions are arbitrary and only the law connecting the initial conditions to the final outcome has an intrinsic meaning. But this arbitrariness of initial conditions corresponds to highly idealized situations in which we can indeed prepare initial conditions according to our will. In complex systems, be it a liquid or, even more so, some social situation, the initial conditions are no longer submitted to our arbitrariness, but are themselves the outcome of the previous evolution of the system.

This connection can be made more explicit through the conceptual framework developed in this paper. Let us compare the distribution function ρ and $\tilde{\rho}$ expanded in terms of the eigenfunctions of the internal time operator. I have already emphasized that in ρ future and past enter symmetrically. Moreover, this symmetry is propagated by the unitary transformation. The situation changes radically when we consider the formula for the transformed distribution $\tilde{\rho}$. As the λ_n values decrease with $n \to +\infty$, the contributions of the partitions belonging to the future are "damped." Past and future enter in a *dissymmetrical* fashion: we have here states with temporal "polarization." Such states can only be the outcome of an evolution that itself is temporally polarized and will remain so in the future.

We see, therefore, that states and laws are indeed closely connected. There are *self-preserving* forms of initial conditions. After all, an initial condition corresponds to a time we choose *arbitrarily;* it can have no basic properties that would distinguish it from all other times. Or, in more philosophical terms, we see a close relation between Being and Becoming. Being is in this way associated with states and Becoming with the laws transforming the states.

In the world of thermodynamics there is an essential temporal element both in Being and Becoming. As Herman Weyl has written in his celebrated monograph on *Space, Time and Matter*[10]: the oneness of the sense of Time exists beyond doubt—indeed, it is the most fundamental fact of our perception of Time.

We begin to see that the arrow of time does not oppose humanity to nature. Far from that, it shows the embedding of humanity in the evolutionary universe that we discover at all levels of scientific endeavor.

17.7. Concluding Remarks

Let me conclude by starting with a somewhat personal remark. When I was a young man, I studied with delight both the work of Bergson and, a little later, the theory of relativity, obviously one of the greatest theoretical constructions ever produced. I found it always astonishing that Bergson had such a difficult time understanding Einstein, and also that Einstein remained *nearly* all his life (I emphasize *nearly* for reasons I shall explain later in this section) hostile to the problematics at the core of Bergson's work. It seems to me that the progress in our understanding of time that I have described in this essay permits us to avoid some aspects of this conflictive situation.

I have described internal time, which exists only for unstable dynamical systems, and shown that its average $<T>$ keeps track with the dynamical time t. However, let me emphasize once more that it is essential not to confuse T and t. The internal time T appears as a kind of "pretime" whose eigenvalues generate the usual astronomical time as read on a watch. I have already in this essay mentioned Aristotle's view that time requires "a soul which counts." This, of course, is outside the range of physics. However, the internal time corresponds precisely to a qualitative change, which is then expressed by "ordinary" time. In this perspective a prerequisite for the use of a "watch time" t is that it could be generated by internal time T.

As is well known, general relativity is based on the four-dimensional interval ds^2. But the specific space-time coordinates that describe this interval are considered arbitrary. A natural supplementary requirement is that the time coordinate t is such that, using *this* time t, the entropy increases. A simple example has been studied recently by Lockhart, Misra, and Prigogine,[11] who demonstrated that for a cosmological model having spatial hypersurfaces of negative curvature it is possible to introduce an internal time closely related to the usual cosmic time. But that is not true in general. For example, in a celebrated cosmological model elaborated by Gödel,[12] an observer always following the direction of increasing time can reenter his or her own past.

In his comment on Gödel's paper, Einstein voices doubts that Gödel's atemporal universe could correspond to the universe in which we live. He writes that "We cannot telegraph into our time," adding:

What is essential in this is the fact that sending of a signal is, in the sense of thermodynamics, an irreversible process, a process which is connected with the growth of entropy (*whereas according to our present knowledge,* all elementary processes are reversible).[13]

It is very interesting to note here that Einstein could not avoid considering irreversibility as an essential ingredient of our world picture.

In short, from the point of view developed in this paper, the difficulty with Gödel's universe is that time as it appears there cannot derive from any internal time related to entropy increase and can therefore not be used as a reference for any observer who would be subject to the second law of thermodynamics. The second law of thermodynamics plays the role of a selection principle for space and time concepts as used in general relativity. In a paper to appear shortly, Géhéniau and I will explore the consequences of this new selection principle. I cannot go into details about it here, but I want to mention that in a most interesting paper Misner came to qualitatively similar conclusions.[14] The time problem cannot be discussed only on the basis of general relativity. But this was exactly the conclusion Bergson reached in his *Duration and Simultaneity,*[15] when he wrote that relativity cannot express *"toute la réalité."* However, Bergson confused the issue by stating that "lived time" had to remain outside the reach of physics. In this sense progress can now be made by comparing astronomical time not to lived time, which is a concept difficult to make precise, but to internal time, which is related to dissipation and is therefore at the basis of the very possibility of life.

The great intellectual revolution that led in the sixteenth and seventeenth centuries "from the closed world to the infinite universe," to use the expression of Alexander Koyré,[16] has led finally to a dualism, to a deep schism exemplified in the contemporary world by the work of Samuel Beckett or the paintings of Malevich. In the terms of Koyré:

> The infinite Universe of the New Cosmology, infinite in Duration as well as in Extension, in which eternal matter in accordance with eternal and necessary laws moves endlessly and aimlessly in eternal space, inherited all the ontological attributes of Divinity. Yet only those—all the others the departed God took away with Him.

Present-day physics is rediscovering time, not the old time, according to which the watch is eternally going to its own past, but an internal time, which corresponds to activity and finally to creative processes as envisaged by Whitehead. It is this rediscovery of time which makes, I believe, our period as exciting and full of promise for the future as the great period between Galileo and Newton.

Notes

1. See I. Prigogine and I. Stengers, *Order Out of Chaos* (New York: Bantam Books, 1984).

2. M. Gardner, *The Ambidextrous Universe: Mirror Asymmetry and Time-Reversed Worlds,* 2d ed. (New York: Charles Scribner's Sons, 1979).

3. See I. Prigogine, *From Being to Becoming* (San Francisco: W. H. Freeman Co., 1980); more recent references may be found in B. Misra and I. Prigogine, "Irreversibility and Nonlocality," *Letters in Mathematical Physics* 7 (1983):421–29.

4. R. C. Tolman, *The Principles of Statistical Mechanics* (London: Oxford University Press, 1938).

5. C. George, F. Mayné, and I. Prigogine, "Scattering Theory in Superspace," *Advances in Chemical Physics* (to appear in 1985).

6. See note 3.

7. See D. N. Parks and N. J. Thrist, *Times, Spaces and Places: A Chrono-geographic Perspective* (New York: John Wiley Sons, 1980).

8. H. Bergson, *Oeuvres,* Editions du Centenaire (Paris: PUF, 1970).

9. A. N. Whitehead, *Process and Reality: An Essay in Cosmology* (New York: Free Press, 1978).

10. H. Weyl, *Space, Time, Matter,* 4th ed. (1922), trans. Henry L. Brose (Dover Publications, 1950).

11. C. M. Lockhart, B. Misra and I. Prigogine, "Geodesic Instability and Internal Time in Relativistic Cosmology," *Physical Review D* 25, 1 (1982):921–29.

12. K. Gödel, "An Example of a New Type of Cosmological Solutions of Einstein's Field Equations of Gravitation," *Reviews of Modern Physics* 21 (1949):447–50.

13. P. A. Schlipp, *Albert Einstein* (New York: Tudor Publishing Co., 1951); emphasis in original.

14. Charles W. Misner, "Absolute Zero of Time," *Physical Review* 196, (1969):1328–33; I. Prigogine and J. Géhéniau, essay to appear in *PNAS*.

15. H. Bergson, *Duration and Simultaneity,* trans. Leon Jacobson (Indianapolis: Bobbs-Merrill, 1965).

16. Alexandre Koyré, *From the Closed World to the Infinite Universe* (Baltimore and London: Johns Hopkins University Press, 1957).

18. Joseph E. Earley

FAR-FROM-EQUILIBRIUM THERMODYNAMICS
AND PROCESS THOUGHT

In previous essays (Earley 1981a, 1981b, 1984) I have attempted to relate some recent scientific advances to the main questions of process philosophy. However, I believe that considerable caution must be used in drawing analogies between physical and philosophical questions. For instance, the important concept of "nonlocality" discussed in Ilya Prigogine's work should not be interpreted in a way that might lead to violation of Whitehead's "ontological principle." Prigogine's work provides no basis for ascribing effects to causes unconnected in time or space with the result caused, no warrant for a "morphogenetic field."

The prominent chemists R. E. Connick and B. J. Alder (1983) carried out a detailed computational study of a type of "elementary reaction" in which one water molecule replaces another in the immediate vicinity of a metal ion in aqueous solution. A specific example of the general class of reaction under study is:

$$Ti(H_2O)_6{}^{3+} + H_2O' \rightarrow (H_2O')Ti(H_2O)_5{}^{3+} + H_2O.$$

So far as chemical reactions in water solution are concerned this one is simple in the extreme, yet even here one must take more than thirty water molecules into account to understand the main features of the process. Change occurs only by concerted movement of over one hundred atoms. There is nothing here *apart from* motions of individual atoms, but there is definitely something important that is *other than* few-body interactions and states of space-time of atomic or subatomic dimensions. The point is no more (and no less) than the metaphysical question whether the problem is solved at the level of few-body interactions or at a "chemical" level. The region of space-time that must be taken into account to understand this "elementary" reaction includes many regions of the size that need to be considered to solve other problems. Related results have been obtained in studies (Hopfield, 1984) of the behavior of large arrays of simple computing devices (called "neurons"); "emergent collective properties" exist that are nearly independent (within wide limits) of the detailed nature of the individual devices. It seems to me that this is what Prigogine means by "non-localized." No problem is *apart from* locality; rather, the region that is important is *other than* a small one, such as might be occupied by a single "elementary particle."

Let me balance my caution against overinterpretation of "nonlocality" with an enthusiastic recommendation of Prigogine's discussion involving the "baker transformation." Even though most readers may not recognize it at once, that discussion is most useful in clarifying Whitehead's ninth Categoreal Obligation, "The Category of Freedom and Determination" (PR 27). Briefly, this states that "in each concrescence whatever is determinable is determined, but . . . there is always a remainder for the decision of the subject-superject of the concrescence." Contemporary work in chemical and physical dynamics, in mathematics, and in meteorology shows that systems governed by fully deterministic laws still can yield outcomes that are not predictable from the governing relationships and the initial conditions alone. The current scientific ferment in this area should be attended to by philosophers.

The main claim I would like to make is that Prigogine's notion of "dissipative structure" is important for one of the principal problems of process philosophy, what Charles Hartshorne named "the problem of the compound individual." The ubiquity of the compound-individual problem is indicated by Whitehead: "However we fix a determinate entity there is always a narrower determination of something that is presupposed by our first choice. Also there is also a wider determination into which our first choice fades by transition beyond itself. . . . Thus no one word can be adequate. But conversely, nothing must be left out" (SMW 93).

How could a person distinguish whether a compound individual were one or many? In terms of the "pragmatic maxim" of C. S. Peirce: "What effects, that conceivably have practical bearings," depend on how the problem of the compound individual is resolved? Are there effects that differ according to whether a compound individual is one or is many?

Prigogine has made the point that the concept of the "simplicity" of the ultramicroscopic belongs to the physics of the past, not to the physics of the present. There is, in his view, no "fundamental" level of description. This doctrine entails corollaries. There is no level of actual entity that is, univocally, fundamental. There are no "elementary" entities; every thing is, in some sense, composite. Every one is, somehow, many. No entity is simply a unit. There are no "simple" physical prehensions; all prehensions are "transmuted."

If we take the position that there is no one level (spatial or temporal) of entity that is noncomposite, to what kind of ontology, what sort of epistemology, does this lead? If no *level* is fundamental, is there some *kind* of interrelation of parts that gives rise to unification that merits special (ontological or even metaphysical) status?

In the *Sophist* (246–47), Plato discussed "a sort of war of Giants and Gods" that was then going on between "those who make being to consist in ideas . . . they are civil people enough" and those "who define being and body as one. . . . terrible fellows they are." To resolve this dispute he suggested a doctrine that is not that of either of the two factions mentioned: "My suggestion would be that anything that posseses any sort of power to affect another, or to be affected by another even for a moment, however trifling the cause and however slight and momentary the affect, has real existence; and I hold that the definition of being is simply power" (cf. AI 119 ff.) According to Whitehead: "whenever the 'all or none' principle operates we are in some way dealing with one actual entity and not with a multiplicity of such entities" (S 28). Is the "all-or-none character" somehow related to the power of which Plato speaks?

What could all this mean in a concrete and particular case? Is the sun metaphysically one or is it, in Leibniz's phrase, a "mere aggregatum"? The sun is an example of a dissipative structure of the type discussed by Prigogine. This new concept includes essential aspects of solar existence that previous notions have underrepresented. In the interior of the sun, hydrogen atoms are converted to helium atoms in an autocatalytic, multistep, cyclical process involving transient formation of oxygen, carbon, and nitrogen atoms (Bethe's carbon cycle). Helium atoms are somewhat lighter than the total weight of four hydrogen atoms. The extra mass shows up as energy released. The cycle of nuclear reactions provides more energy that it consumes. This autocatalytic cycle is opposed and balanced by mutual gravitational attraction of the components of the sun. This network of interactions is quite analogous to the chemical systems discussed on a technical, scientific level by the Brussels and Austin research groups led by Prigogine and mentioned in his present paper. The "Brusselator" and the Belousov-Zhabotinsky reaction (modeled by the "Oregonator") are both dissipative structures.

In each of these cases there is a closure of a cycle of relationships (chemical reactions). Such a closure is an all-or-none phenomenon. In the case of the sun this closed network of reactions offsets the gravitational force tending to crush the components together. The heat production of the cycle generates high local temperatures: the sun radiates energy to the rest of the universe. The result is homeostasis that necessarily involves outward flux of matter and energy. That flux links the sun with other entities, including us. The sun is a structure, but one that necessarily involves net production of entropy and con-comitant interaction with the rest of the universe. The balance of the two opposing sets of processes is neither perfect nor indefinetly constant. Solar hydrogen is being used up and will in the distant future be exhausted, but the balance between the gravitational attractions and the energy-producing cycle of nuclear reactions is sufficiently close and constant that terrestrial life has been able to evolve, powered by solar radiation that has been quite stable for millions of years. So far as "effects that have practical consequences" the sun is "a thing" one thing. The sun has as much claim to be considered, metaphysically, as one entity as has any other candidate for that status (any human person, any minuscule space-time region).

Homeostatic interaction networks exist at many levels of scale, from a combination of minuscule quarks that constitute a tiny proton (in-volved in coulombic interactions with other charged particles) to an aggregation of cold clouds of diffuse gas that form a hot star radiating across a wide spectrum. Many aggregations, many networks of processes, approximate homeostasis only roughly. For most, the approach is not sufficiently close as to be worth discusssing. But some aggregations approach homeostasis well enough to achieve metaphysically significant unity.

Compound individuals persist when their internal composition and external circumstances conspire to produce a unique series of states that continually repeat, in spite of fluctuations and variations. The persistence of the unique sequence of states depends on the closure of a particular kind of network of internal relationships (that define a homeostatic cycle). Persistence of the structure also depends on the appropriateness and regularity of sustenance received from other struc-tures both outside and within the compound individual.

Regularity of external environment and dependable functioning of interior structures require closure of many networks of interaction (both larger and smaller than the interaction cycles of primary interest). "The problem of evolution is the development of new shapes of value that merge into higher attainments of things beyond themselves" (SMW 93). That problem has been provisionally solved (new shapes of value have been developed) as many times as novel kinds of entities have evolved, and also as many times as an instantiation of any one of those kinds has become realized. Each of these provisional solutions has built upon prior closures, and each has, by its valuable achievement, increased the scope for subsequent integration.

My thesis is that an entity should be regarded, metaphysically, as one entity if, and to the extent that, it interacts with particular percipients as unified. Ontology and epistemology are inextricably interrelated. "Causal efficacy" is governed by transmutations that should be understood as components of the epistemological (broadly understood) aspect of the entity. "Presentational immediacy" may be, but need not be, involved. Unifications of submicroscopic entities of particle physics that yield stable aggregates, combinations of genes that correspond to viable organisms, networks of behavior patterns that specify evolutionarily stable strategies, all of these might form the basis of metaphysically significant unicity.

Each of these types of unity corresponds to a type of dissipative structure as discussed by Prigogine. It seems to me that the far-from-equilibrium structures discussed by Prigogine provide a more appropriate cosmological framework for process thought than does any previous type of scientific discourse.

Abbreviations of Whitehead's Works

AI   *Adventures of Ideas* (New York: Free Press, 1967)
PR   *Process and Reality,* Corrected edition, ed. David Ray Griffin and Donald W. Sherburne (New York: Free Press, 1978)
S   *Symbolism: Its Meaning and Effect* (New York: Capricorn Books, 1927)
SMW *Science and the Modern World* (New York: Free Press, 1967)

References

Connick, R. E., and Alder, B. J. 1983. "Computer Modeling of Rare Solvent Exchange." *Journal of Physical Chemistry* 87:2764–71

Earley, J. E. 1981a. "On Applying Whitehead's First Category of Existence." *Process Studies* 11, 1:35–39.

———. 1981b. "Self-Organization and Agency: In Chemistry and in Process Philosophy." *Process Studies* 11, 4:241–58.

———. 1984. "Evolutionary Biology, Physiological Psychology and Process Philosophy." Paper delivered to the Society for the Study of Process Philosophies held in conjunction with the meeting of the Metaphysical Society of America, Evanston, Illinois, March.

Hopfield, J. J. 1984. "Neurons with Graded Response Have Collective Computional Properties Like Those of Two-State Neurons." *Proceedings of the National Academy of Sciences, U.S.A.* 81:3088–94.

19. Pete A. Y. Gunter

SOME QUESTIONS FOR ILYA PRIGOGINE

The work of Ilya Prigogine and his colleagues strikes me—and many others—as some of the most fascinating and potentially fruitful going on in science today. It has therefore come as a surprise to me to encounter skepticism from many scientific colleagues concerning nonlinear nonequilibrium thermodynamics. I have strong suspicions that the greater part of this skepticism stems from a simple unwillingness to "take time seriously," as Samuel Alexander put it: to think in process terms. As Henri Bergson knew, it costs a real effort to do so. But there may be another factor involved, namely, theoretical or experimental incompleteness.

My first question concerns thermodynamics and the unity of science. It is divided into three parts. Nonequilibrium thermodynamics might be incomplete with regard to three disciplines: thermodynamics *per se,* the rest of physics, or biology (including biochemistry). I will deal with each of these in turn.

Nonequilibrium thermodynamics rests on two symmetry-breaking operations. One of these establishes temporal irreversibility; the other establishes a "preferred" (i.e., one-way) direction of time. Do either of these operations, and the experiments on which they are based and which they are required to explain, require either to be more firmly or more broadly established? Is it necessary to extend them to hitherto unexplored domains, to add experimental proof or mathematical formalisms?

With regard to the other "parts" of physics, the questions may appear more familiar. Karl Popper has written a book entitled *The Schism in Physics.* It seems to me that he is too optimistic. Physics is at present divided into three parts: quantum physics, relativity physics, and thermodynamics—four parts, if cosmology can be treated as independent. How, one wonders, does Prigogine see the relations between these disciplines and nonlinear thermodynamics? Will the new thermodynamics have to reformulate or even absorb quantum or relativity physics before doubt is erased in some minds?

Finally, there is biology. Bénard instabilities, Leisegang rings, and Belousov-Zhabotinsky reactions are a long way from living cells, DNA, or even the full panoply of amino acids necessary to create life. Limit cycles, biochemical rhythms of all kinds, seem less so, particularly when coupled with temporal conceptions of cell structure and intercellular communication. One wonders how Prigogine sees the relations between his new thermodynamics and the life sciences. Have we far to go before biology and biochemistry become components of this new physics?

Next, I have two questions about Prigogine's personal and intellectual background. First, in his autobiographical notes, Einstein mentions coming to the basic notion of relativity through a paradox he hit upon at the age of sixteen, according to which if one traveled at the speed of light alongside a moving light ray, it ought to appear as a standing wave. No such thing appeared possible to Einstein, who set out to question the assumptions on which the impossibility was based. I wonder, analogously, whether Prigogine's initial model was arrived at experientially, or from the study of philosophy or art?

Second, Prigogine has mentioned the divergence between Bergson and Einstein: Bergson with his irreversible, qualitative duration, Einstein with his geometrical space-time. Would Prigogine suggest to us the manner in which he initially conceived these two thinkers, as well as the way in which his work may be considered to reconcile their views toward space and time?

Finally, I have two questions about physics. The first of these is whether Prigogine suspects that the microcosm is the true source of irreversibility. The second runs as follows: Are we sure that quantum physics is deterministic? Prigogine has said that the evolution of quantum physics' wave equation is deterministic, and yet quantum physics is, as everyone knows, intrinsically stochastic. How can a science be both probabilistic and deterministic?

20. Ilya Prigogine

RESPONSE TO PETE GUNTER

We all know that we are going through an age of transition. It is therefore natural that we explore various directions and that no consensus has yet been reached. However, it seems to me that the role of time, of evolution, and therefore also of irreversibility and entropy is steadily increasing in importance. Moreover, I believe that irreversibility will indeed play an unifying role.

As for the first point, nonequilibrium physics and chemistry are now well-defined fields of inquiry, studied both from theoretical and experimental points of view, in a number of laboratories, both in Europe and the United States. We know that, far from equilibrium, new structures may arise, be they "regular," as in limit cycles, or "chaotic," as in the case of strange attractors. Much remains to be done, but recent progress has been spectacular and is likely to continue for several decades.

The second point raised by Gunter is the relation between thermodynamics, dealing with irreversibility, and the other disciplines of physics. One must remember that, in the classical view, irreversibility is added through supplementary approximations. This hardly corresponds to the basic role of irreversibility, as implied by the recent developments I just mentioned. For this reason we now start with the second law considered as a dynamical principle. This leads to a theory of transformation between groups and semigroups and therefore between deterministic and probabilistic processes such as are exemplified by the Markov chains. In simple situations (such as K-flows or potential scattering), the theory seems to me pretty satisfactory. However, much remains to be done. Computer experiments are now been performed to test our microscopic theory of irreversible processes.

I hope that careful experiments on the interaction of matter and light may lead to interesting conclusions in this respect. I do not believe that thermodynamics will absorb other disciplines, such as quantum or relativistic physics, but I do believe that these disciplines must be reformulated in order to absorb the concept of irreversibility.

Irreversibility is indeed such a general concept that it may arise at various levels, be it classical, quantum, or relativistic. Think, for example, about a gas of hard spheres. Here irreversibility arises in a system that may first be described as a classical dynamical system. Consider, in contrast, chemical reactions that always lead to irreversibility; kinetic constants can be related to quantum physics; therefore, in this sense, the origin of irreversibility lies in quantum effects. Still, no *explicit* use of quantum mechanics is necessary to describe the production of entropy in chemical systems.

In traditional quantum mechanics, irreversibility arises only through measurement and observations. This qualification leads to views such as the participatory universe of J. A. Wheeler, in which it is the observer who creates the history of the universe. Wheeler's view contrasts with my approach, which takes account of the fact that there are classes of physical processes that are irreversible independently of any human act of measurement or observation. The "participatory universe" now has a different meaning. The phrase expresses the fact that human thought and measurement is one part of the effectiveness of irreversibility as printed in the physical world.

Let us now turn to biology. I believe there is now a growing consensus that irreversibility may play a role in many mechanisms characteristic of life, such as regulation and recognition. I have already mentioned the fact that nonequilibrium structures may be "regular" or chaotic." From this point of view, it is interesting to note that cardiac diseases are related to rhythmic disorder; and some neurological troubles seem to correspond to an excess of order as recorded in EEG.

The question of the incidence of irreversibility in the origin of life still seems a very controversial one. I believe we can safely hold that the well-known paradox of "zero probability" for biomolecules in an equilibrium environment is solved by recognizing that the nonequilibrium conditions prevailing in the earliest ages of our planet restricted the possibilities very severely. Consider the Bénard instability, of which the probability also tends to zero near equilibrium.

The precise way in which irreversibility enters into biology is now understood in simple cases such as enzymatic chemistry.[1] More generally, biology has a basic historical component, as everyone has known since Darwin. The origin of life cannot be treated independently of the transcription of irreversibility at the molecular level. Considering the structure of a crystal, we may soundly infer the nonequilibrium conditions in which it grew. We may hope to do the same for the nonequilibrium conditions of biomolecules, but this program is still just beginning. One of the most interesting problems here, closely related to the origin of life, is the inscription of nonequilibrium conditions in polymer material.

The last question asked by Gunter refers to terminology. Quantum theory is deterministic if one considers the wave functions and probabilistic if one considers the distribution of observables. This duality corresponds to the traditional attitude and is generally accepted. However, one should not confuse a probabilistic theory, such as quantum theory, and the theory of stochastic processes, such as Markov chains associated with an increasing entropy function.

In short, it seems to me that the present situation in physics calls for a reformulation of the basic laws in terms of appropriate semigroups. The main point is, of course, that in the theoretical description of dynamical groups past and future play the same role, whereas this is no longer the case with semigroups.

Note

1. See G. Nicolis and I. Prigogine, *Self-Organization in Non-Equilibrium Systems* (New York: Wiley, 1977).

21. David Bohm

COMMENTS ON ILYA PRIGOGINE'S PROGRAM

Ilya Prigogine has proposed, in "Irreversibility and Space-Time Structure," that the study of complex systems (including, for example, biology and the human sciences) may reveal features of natural law that are just as fundamental as are those disclosed by physics and chemistry in the study of simple systems. This is a revolutionary suggestion, with which I am in complete agreement. It is an important step toward wholeness in our thinking, insofar as it may lead us to see that the divisions between "large" and "small," and between "simple" and "complex," are of only a relative and limited kind of significance.

Prigogine and his co-workers have made an extensive study of non-equilibrium processes in complex dissipative systems, such as gases or liquids (perhaps also undergoing simple chemical processes), in which even relatively simple laws allow for the rise of complex and qualitatively new kinds of behavior. In this behavior, temporal and spatial symmetry is broken and a rich diversity of space and time patterns is displayed. Many of these patterns have in fact been confirmed experimentally.

The main point discussed in Prigogine's paper is how such dissipative behavior is to be understood from the point of view of the fundamental laws of physics, which are reversible and which therefore would seem to have no place in them for phenomena based on the actually observed irreversible nature of large-scale physical process. Prigogine and his co-workers show that this behavior can be explained on the basis of a certain kind of instability in the motions of a wide variety of systems. For such unstable systems, a transformation is possible that changes the *reversible* basic equations of classical physics into a form similar to that of *irreversible* (dissipative) systems. Prigogine states that this transformation amounts to a choice of just those critical conditions that will eventually bring about a steady equilibrium state. The proof that this transformation can be made will constitute a demonstration of the compatibility of the reversible classical laws with the actually observed irreversible nature of all large-scale processes.

In formulating the transformation described above, it is necessary to introduce the notion of *internal time*. That is, the passage of time may be related to complex changes in the internal state of a system that correspond roughly to the idea of the "aging" of that system. Mathematically, internal time can be expressed in terms of an operator on the distribution function. Prigogine is suggesting that this notion of time (as "age") is more fundamental than is the notion of time as measured by the motion of an essentially reversible system (e.g., the planets) or by a watch.

Prigogine's suggestion of the primacy of internal time is very interesting and deserves a great deal of further development. However, it requires conceptual clarification in a number of ways. In particular, it is completely ambiguous what is to be meant by the "simultaneous existence" of systems of different ages. Leibniz has defined space as the order of coexistence and time as the order of successive existence. In his paper, Prigogine has emphasized the order of successive existence and has not discussed what is to be meant by coexistence.

Prigogine does, however, introduce a distinction between "age time" and "ordinary time" (or "watch time"). The latter is ruled by the laws of relativity theory (special and general). In relativity theory the notion of coexistence is modified, in the sense that whatever is outside the light cone of a given event may be said to coexist with it. But it is necessary to note that this crucial feature of our time concept is not contained as an intrinsic feature of "age time." Whenever two systems have to be treated together, it therefore appears that we will have to appeal to two kinds of time, i.e., "watch time" and "age time." These are, of course, related by the laws of physics. Thus, with the aid of cosmological models, Prigogine shows that in certain cases changes in "watch time" are *quantitatively* derivable from changes in "age time." Nevertheless, the qualitative feature of coexistence cannot be derived in this way, and so there is still the need to introduce two different (though related) kinds of time. It would be extremely interesting to explore the possibility of deriving even the feature of coexistence from age time.

This question is perhaps related to another one that puzzles me on reading Prigogine's paper. In this paper it appears that each classical system is well defined in its state (so that its movement corresponds to a well-defined trajectory in phase space). Yet, in the theory, it appears that each system is also a (perhaps small) region in phase space. It is this region that diffuses and approaches a uniform equilibrium state. And yet a given individual system does not actually diffuse, but always remains well defined. It seems to me that Prigogine in this way is tacitly treating a given system as being, in some sense, inherently ambiguous. I would have no objection to this in principle; indeed, I think that this would be an interesting new idea, in the sense that classical and quantum mechanics would share the notion of a certain inherent ambiguity in the basic properties of matter. However, I think that it would be helpful if this point could be clarified and made explicit.

Finally, I would like to suggest a certain relationship between Prigogine's ideas and my own notions on the implicate order. To bring this out I would like to add that (as pointed out in my paper) two key aspects underlie the concept of time. These are *recurrent process* and *nonrecurrent process*. Mathematically, recurrent process (e.g., oscillation) is treated most naturally in terms of Fourier series and therefore of exponentials with imaginary arguments (such as $e^{i\omega t}$). The most common kind of nonrecurrent process is growth and decay, which is treated in terms of exponentials with real arguments ($e^{\lambda t}$ and $e^{-\lambda t}$).

Now, quantum mechanics is basically tied to exponentials with imaginary arguments (because of the reversible character of Schrödinger's equation). On the other hand, the irreversible processes discussed by Prigogine are treated through the use of exponentials with real arguments. My suggestion, then, is that if we generalize to exponentials with complex arguments ($e^{(-\lambda + i\omega)t}$), we should be able to comprehend both recurrent (reversible) and nonrecurrent (irreversible) processes within one framework of concepts. This would, in effect, extend the implicate order such that it would imply a fundamental relationship between quantum mechanics and irreversible processes of the kind discussed by Prigogine. Such a unification of quantum processes and irreversible processes would thus be of a fundamental nature, not contained in presently accepted theories. It would therefore go beyond Prigogine's further proposal, in which the approach described in his paper is to be applied to the probability distributions of the current quantum theory rather than to those of the classical theory. I am at present looking at theories based on the suggestion outlined above, though in a preliminary way, and I hope that there will soon be some progress in this direction.

22. Henry P. Stapp

EINSTEIN TIME AND PROCESS TIME

To comprehend the significance of time in modern physics one must distinguish two very different kinds of time. The first I call "process time," the second "Einstein time." Process time is the time associated with a cumulative process whereby things gradually become fixed and settled. Einstein time is the time part of the space-time continuum of contemporary physics.

Contemporary physical theory establishes no connection at all between these two kinds of time, for it says nothing about process. It deals rather with the content of observations. Each observation has a content that includes, in principle, a clock and ruler reading. These readings assign to the observation a place in the space-time continuum. But whether the data represented in one observation become fixed and settled before or after the data represented in some other observation is not determined by contemporary physics: one can equally well imagine *either* that everything becomes fixed and settled all at once, in some single act of creation, and hence that neither process nor process time exists, *or,* alternatively, that things become fixed and settled in some definite order. These two possibilities are not empirically distinguishable. Indeed, Einstein's analysis of the meaning of time in physics made it clear that time enters physics only through the content of observations that say nothing at all about the order in which things become fixed and settled. His analysis effectively banished the concept of process from the physical theory of his era.

Of course, in the deterministic framework within which Einstein himself worked, process could be no real issue. The deterministic laws ensured that everything was fixed and settled by the initial act of creation. Thus there could be no process. Hence the real impact of Einstein's analysis of time came only later, when quantum theory introduced indeterminism. In this latter context the idea of process arises naturally, at least at the conceptual level. But the founders of quantum theory, following Einstein's lead, circumvented the problems associated with process by asserting that the quantum-mechanical formalism "merely offers rules of calculation for the deduction of expectations about observations obtained under well-defined experimental conditions specified by classical physical concepts."[1]

Quantum theory has, nevertheless, one feature that suggests that it should be formulated as a theory of process. The wave function of the quantum theory is most naturally interpreted as representing "tendencies" or "potentia" for actual events. This intuitive idea of the meaning of the wave function was first made explicit by David Bohm in his 1951 textbook, *Quantum Theory.*[2] The idea was endorsed by Heisenberg[3] and probably agrees with the intuitive ideas of most quantum physicists. However, in spite of this natural occurrence within quantum theory of this basic process notion of "tendency," the theory was not originally formulated as a theory of process. Severe technical difficulties blocked the way.

Of the many difficulties blocking the development of a quantum theory of process I shall mention only two. The first is that the world at the observed macroscopic level conforms generally to the concepts of classical physics, whereas the world at the atomic level is represented by a very different kind of mathematical structure. In particular, at any instant of time a system of n particles is represented in quantum mechanics by a function in a space of $3n$ dimensions, whereas the world of classical physics resides in a space of 3 dimensions. There appeared to be no mathematically and physically acceptable way to relate in any exact way the quantum structure in $3n$ dimensions to classical physics in 3 dimensions, without referring explicitly to "our observations" and exploiting the empirical fact that we describe the content of our observations by using classical concepts. This procedure was considered unsatisfactory by some scientists and philosophers because it injects human observers into the structure of basic physical theory. Their objection was met by insisting that science should deal directly with what humans can know or expect. This second point of view has prevailed partly, at least, because no natural process formulation of quantum theory has appeared.

A second difficulty was that the identification of the "actual events" appeared to require the notion of irreversible processes, and this notion seemed ill defined on a fundamental level.

Recent technical advances appear to have opened the way to the development of a natural quantum theory of process. The two key ingredients are the S-matrix and some recent developments in the quantum theory of light.

The S-matrix was introduced into quantum theory by Wheeler and was used by Heisenberg to get around a serious technical problem. This problem had its conceptual side. Relativistic quantum theory had taken over from nonrelativistic quantum theory the notion of the state of a system existing over all of space at a single instant of time. This concept does not conform to the relativistic precept of the meaninglessness of simultaneity. On the other hand, on the practical side, the introduction of the requirements of the theory of relativity had engendered certain divergences, called "ultraviolet" divergences, that caused the state of the system at an instant of time to become mathematically ill defined.

Heisenberg observed that both problems could be resolved, without losing any of the empirical content of relativistic quantum theory, by abandoning the concept of the state of the system at an instant of time and adopting, instead, the idea that the basic quantity in relativistic quantum theory was the S-matrix.

The S-matrix viewpoint would appear at first to eliminate completely any possibility of developing a quantum theory of process, for the S-matrix describes, by definition, only the quantum transitions from states in the infinite past to the states in the infinite future. It omits all description of what is going on in between. Moreover, space is also eliminated. For the basic variables of the S-matrix are not the variables that represent position, but rather variables that represent states of inertia—states that if left undisturbed would endure unchanging for all eternity.

The S-matrix viewpoint appeared beset also by a serious technical difficulty: the S-matrix seemed to be mathematically well defined only in an approximation where light was ignored. This was the famous "infrared catastrophe." Yet light certainly cannot be ignored in any adequate description of the world in which we live, for light (i.e., the electromagnetic interaction) is responsible for the binding that holds together both the objects of everyday experience and our bodies themselves. Light is the light by which we see, it is responsible for the sound by which we hear, and so on.

A completely satisfactory resolution of this infrared catastrophe has only recently been found.[4] It is based on the discovery that the light emitted by matter has a well-defined classical part. This classical part is a function defined over the four-dimensional space-time continuum. Thus the incorporation of light into the S-matrix—and light is the very ingredient most responsible for the form of the world in which we live—automatically brings into the S-matrix formulation of relativistic quantum mechanics an *exact* classical level of description coordinated to the ordinary four-dimensional space-time continuum of the theory of relativity.

This technical development provides the foundation for a natural formulation of a quantum theory of process. Process is considered to consist of a well-ordered sequence of actual events. Just prior (in process time) to event number n there are several possibilities for what might become actualized. Let these possibilities be labeled by the set E_n of indices e_n. One of these events will become actualized. Let it be called a_n.

For each n there is a corresponding space-time region R_n, determined by process. For each possible event e_n there is a projection operator $P(e_n)$ that restricts the classical field inside region R_n to some subset of the set of all conceivable classical fields in R_n.

Let $S(e_n)$ represent the S-matrix subjected to the condition that the classical fields be limited to those allowed by the product of projection operators $P(e_n)P(a_{n-1})P(a_{n-2})$ . . . . Let d_{in} be the initial density matrix, and let $d(e_n)$ be $S(e_n)d_{in}S^+(e_n)$. Then the probability for event number n to be e_n is

$$\frac{\text{tr } [d(e_n)]}{\Sigma \text{ tr } [d(e'_n)]}$$

where the sum appearing in the denominator is over all e'_n in E_n.

The set E_n of possible events e_n is restricted by the following bifurcation condition, which will be discussed later: the various operators $d(e_n)$ for e_n in E_n have disjoint supports in the multiparticle momentum space upon which they are defined.

The process formulation of quantum theory contains no explicit dependence on human observers: it allows quantum theory to be regarded as a theory describing the actual unfolding or development of the universe itself, rather than merely a tool by which scientists can, under special conditions, form expectations about their observations. The quantum theory of process is in general in accord with the ideas of the physicist, logician, and process philosopher Alfred North Whitehead.[5] In particular, the actual is represented not by an advancing, infinitely thin slice through the space-time continuum, but rather by a sequence of actual becomings, each of which refers to a bounded space-time region: event number n is represented, within physical theory, by a restriction on the set of classical fields allowed in the bounded space-time region R_n. We have, therefore, neither becoming in three-dimensional space nor being in the four-dimensional world,[6] but rather becoming in the four-dimensional world.

Event number n is represented in physical theory by a restriction upon the classical fields allowed in the bounded space-time region R_n. This restriction induces, through the quantum formalism, changes in the tendencies for the next event. These changes in tendencies are manifested over all of space-time—i.e., even if the region R_{n+1} is spacelike situated relative to R_n. This change in tendencies is the nonlocal change that is associated with the collapse of the wave function in some formalisms and, more generally, with Bell's theorem.[7]

The tendencies are calculated in the quantum formalism by using Feynman's sum over all space-time paths. In the S-matrix formulation, these paths extend in time, in general, from minus infinity to plus infinity: they do not terminate in the region R_n. Thus as regards tendencies the entire space-time continuum of relativity theory is involved in each step of the process of becoming. But as regards actualities each actual event is associated with a bounded region in space-time.

The conception of process described above differs from Whitehead's because it has no place for his "contemporary events." However, because Whitehead stressed so often that these awkward contemporary events were forced upon him by the physics of his time, rather than by his general principles, it is, I think, safe to infer that, had he known about the nonlocal connections entailed by Bell's theorem, he never would have mutilated his theory by the introduction of these contemporary events.

A brief remark about gravitation is needed. The gravitational field is a gauge field that can be expected to have a classical part similar to the one associated with light. But the details have not yet been worked out. So I have simulated here the expanding universe in which we live by a flat space and an initial state in which all constituents are rushing toward the space-time region of the big bang. This allows S-matrix ideas to be used. The final state can be conceived as a state where the various momenta are all quite accurately defined. The partially neglected gravitational interaction will then turn the outgoing particles into the incoming particles for the next big bang, etc. I have neglected, therefore, both the very long-range part of the gravitational interaction and also any short-range parts that cannot be adequately approximated by gravitational interactions in a flat space.

I turn now to the questions of irreversibility and the connection of the present work to the one presented to this conference by Prigogine. The traditional problem as regards the identification of actual quantum events is that one does not know, with certainty, whether the event is irreversible until one knows exactly what conditions the resultants of the event will eventually encounter: the conditions might conceivably be exactly right to effect a coherent recombination of these resultants. This problem is resolved here by using the overall space-time viewpoint of the theory of relativity and of S-matrix theory and calculating all probabilities from the asymptotic state corresponding to time $= +\infty$. The intuitive idea of physicists is that once the differences between the several possible results have evolved to a level where the different possibilities correspond to different motions of macroscopic objects, or to different macroscopic electromagnetic fields, then the subsequent time evolution will not be able to bring *all* of the 10^{23} momenta corresponding to the two alternative possible results back into coincidence. This is because the interaction required to bring the various parts back together will transfer momentum to other objects, etc. This normal intuitive idea of the condition that defines an actual event is formulated here as the bifurcation condition, which demands disjunction at time $= + \infty$ of the momenta associated with the alternative possibilities.

Although this bifurcation condition corresponds to normal intuitive ideas about the conditions needed to define an actual quantum event, the question arises to what extent this condition can be analyzed with mathematical precision. Here it appears that the mathematical methods employed by Prigogine might prove useful, for in both cases the concern is with the behavior of trajectories when extrapolation is made into the distant future. Prigogine's methods, though dealing in the first instance with classical systems, nevertheless involve probability functions, which in our case follow directly from our use of quantum theory. Conversely, the general framework described here might provide a satisfactory conceptual underpinning for Prigogine's work. His method involves the introduction of a transformation constructed to achieve the desired result of breaking the symmetry between the two directions of time. But the physical origin of this transformation is unclear. In the present framework the symmetry is broken by process itself. Thus the present work, and that of Prigogine and associates, may be two mathematical aspects of a single solution to the problem of understanding process within contemporary physics.

Scientific application of this theory of quantum process may come first in connection with the problem of molecular structure.[8] Indeed, evidence already exists that the assumption needed for the application of the Copenhagen interpretation, namely, that the quantum system can be considered essentially isolated from its environment, fails at the molecular level and that, moreover, the interaction with environmental infrared photons is responsible for the emergence of classical properties at the molecular level.[9]

Acknowledgment

This work was supported by the Director of Energy Research, Office of High Energy and Nuclear Physics, Division of High Energy Physics of the U.S. Department of Energy under Contract DE–AC03–76SF00098.

Notes

1. Niels Bohr, *Essays, 1958–1962, on Atomic Physics and Human Knowledge* (New York: Wiley, 1963), 60; Henry P. Stapp, "The Copenhagen Interpretation," *American Journal of Physics* 40, 8 (1972):1098–1116.

2. David Bohm, *Quantum Theory* (Englewood Cliffs, NJ: Prentice-Hall, 1951).

3. Werner Heisenberg, *Physics and Philosophy* (New York: Harper and Row, 1958), chap. 3.

4. Henry P. Stapp, "Exact Solution of the Infrared Problem," *Physical Review D* 28 (1983):1386–1418.

5. Alfred North Whitehead, *Process and Reality* (New York: Macmillan, 1929); Henry P. Stapp, "Whiteheadian Approach to Quantum Theory and the Generalized Bell's Theorem," *Foundations of Physics* 9, 1–2 (1979):1–25.

6. Allusion is made here to Einstein's famous assertion: "becoming in the three-dimensional space is somehow converted into being in the world of four dimensions." For a discussion of Einstein's vacillating views on this question, see Milič Čapek's essays in: (i) *Bergson and the Evolution of Physics,* ed. P. A. Y. Gunter (Knoxville: University of Tennessee Press, 1969); (ii) *The Voices of Time,* ed. J. T. Fraser (New York: George Braziller, 1966); and (iii) *The Concepts of Space and Time,* Boston Studies in the Philosophy of Science, vol. 22, ed. Milič Čapek (Dordrecht and Boston: D. Reidel Publishing Co., 1976).

7. John S. Bell, "On the Einstein-Podolsky-Rosen Paradox," *Physics* (New York) 1 (1964):195–200. Henry P. Stapp, "Bell's Theorem and the Foundations of Quantum Physics," Lawrence Berkeley Laboratory Report LBL 16482 (1983); also in the *American Journal of Physics* 53 (1985):306–17.

8. R. G. Woolley, "Must a Molecule Have a Shape?" *Journal of the American Chemistry Society* 100 (1978):1073–78; "Quantum Mechanical Aspects of the Molecular Structure Hypothesis," *Israel Journal of Chemistry* 19 (1981):30–46.

9. Peter Pfeiffer, "Chiral Molecules—a Superselection Rule Induced by the Radiation Field Thesis," thesis (Zurich: Eidgenoesche Technische Hochschule, 1980); "A Nonlinear Schrödinger Equation Yielding the Shape of Molecules by Spontaneous Symmetry Breaking," in *Quantum Mechanics in Mathematics, Chemistry and Physics,* ed. Karl E. Gustafson and William P. Reinhardt (New York: Plenum, 1981), 255–69. Hans Primas, *Chemistry, Quantum Mechanics, and Reductionism: Perspectives in Theoretical Chemistry: Lecture Notes in Chemistry* (Berlin: Springer, 1981).

23. Tim Eastman

Process Time and Static Time:
A Response to Henry Stapp

> The laws of nature describe nothing but formal possibilities
> since they are expressed by conceptual language. They are
> made in such a way as to apply to all times.
> ———*Von Weizsäcker* (1973, 647)

This temporal invariance of conceptual frameworks in physical science
referred to by Carl von Weizsäcker complicates our efforts toward
formulating an interpretation of "time." In the same way, the descriptions of physics, theoretical and observational, tend to be neutral with
respect to such interpretations or possible ontological inferences. Whitehead stated that "speculative philosophy is the endeavor to frame a
coherent, logical, necessary system of general ideas in terms of which
every element of our experience can be interpreted." Similar criteria
of logic and coherence are the final tests of our interpretive schema.

As noted above for the concept of "time," our theories or descriptive
frameworks typically do not imply a unique interpretation. Von Weizsäcker (1973, 666) states that, for a quantum theory with a finite-
dimensional Hilbert space,

> multiple quantization would not have to start with a particle in an infinite
> continuous space. I would not have to accept space as given but to construct
> it. The simplest and most abstract construction will begin with a two-fold
> alternative, i.e. with one yes-no decision. Multiple quantization starting there
> would mean to build up the objects in the world as enembles of such decisions.
> The number corresponding to one object would mean the number of decisions
> needed for fully defining that object. Now the quantum-mechanics of an
> ensemble of two-fold alternatives nearly automatically leads to a description
> in a three-dimensional real space and hence in the higher steps to the usual
> field theories in three-dimensional space.

In a similar way, Finkelstein (1973) has introduced discrete counting
of time in which every process is conceived of as a "finite assembly
of elementary processes." However, these descriptions, including Stapp's,
are not unique—indeed, the dominant approaches in relativistic quantum theory involve the use of space and time variables as continuous
(Bjorken and Drell, 1965). In attempts at philosophical interpretation,
we need to be especially cautious of interpretations that are dependent

on a particular descriptive framework. In Stapp's case, I refer to his innovative approach to questions of quantum theory using a particular approach to the S-matrix formalism. However, I consider Stapp's suggestions regarding a quantum theory of process fully consistent with the descriptive content of modern physics. Some interpretations of contemporary physical theory do not refer to process explicitly. Stapp demonstrates that such interpretations are incomplete; numerous other works effectively establish the importance of process time in modern physics (e.g., Čapek, 1984; Prigogine, 1980; Prigogine and Stengers, 1984).

Bjelland (this volume; see also Čapek, 1971) states that Bergson's theory of succession includes durational succession as (a) a dynamic relation, (b) an asymmetrical relation, and (c) an internal relation. Čapek and Bjelland have emphasized how Bergson provides an alternative to the classical atomist position, in which only monadic attributes are used for the primary properties of elementary objects or events. D'Espagnat (1973, 730) describes this classical atomism as the "multitudinist" conception

according to which the ultimate reality—all events and/or microscopic objects, each one of them being endowed with simple properties and being such that the interactions of them all—taken as local and causal—would, combined with chance, give rise to the complexity of appearances.

Referring to recent experiments related to Bell's inequality that appear to provide a direct test of Einstein's assumption of local realism, D'Espagnat argues (1973, 1983) that the multitudinist conception is contradicted by experiment "independently of the formalism." This is an important philosophical statement, which is made stronger by not being dependent on the formalism.

Stapp (1979, 1982) has also done important work on Bell's theorem and related questions. I do not interpret Stapp as supporting a multitudinist conception. It is useful, however, to raise this issue because many others have interpreted Whitehead's event pluralism as effectively a multitudinist conception with the "objects" of classical physics replaced by "events." This interpretation is incorrect because Whitehead's arguments show that any multitudinist view commits the "fallacy of misplaced concreteness." As I will show later, Whitehead utilizes the concept of time as a continuum within the mode of coordinate division in a manner that complements his interpretation of events as exemplifying dynamic, internal relations.

Bergson's theory of succession also includes an asymmetrical relation that, in Bjelland's words (this volume), "entails the irreducibility of the causal order to the static order of logical equivalence and identity." Prior to the advent of quantum theory, the causal order was generally presumed to be reducible to the static order of logical equivalence. The special theory of relativity provides no explicit denial of this assumption, and Grünbaum (1967) has even incorporated this assumption in relativity theory to deny any reality to the present except as a subjective "coming into consciousness." Čapek (1975) states that "nothing in relativistic physics supports this view; on the contrary, at any particular moment, future events are intrinsically unobservable by any conceivable observer." Further, he states that "relativity theory strengthened rather than weakened the objective status of becoming." Apparently, the theory in this case does not provide a description leading to a unique interpretation.

In the general theory of relativity, Einstein links an ontology that emphasizes geometric form with the observational realm of clocks and rulers. Some interpreters have gone on to ascribe ontological priority to geometric form (Graves, 1971). Such interpretations have arisen in part from Einstein's own emphasis on geometric form. However, Einstein cannot be easily classified as an epistemological realist or idealist, since he (1949, 684) wrote that

[The scientist] must appear to the systematic epistemologist as a type of unscrupulous opportunist: he appears as realist insofar as he seeks to describe a world independent of the acts of perception; as idealist insofar as he looks upon the concepts and theories as free inventions of the human spirit (not logically derivable from what is empirically given); as positivist insofar as he considers his concepts and theories justified only to the extent to which they furnish a logical representation of relations among sensory experience.

Einstein (1922, 31) also appears to have rejected the concept of the spatialization of time. Whitehead's own work on the theory of relativity was persistently guided by our direct, immediate experience of nature. Stapp effectively adopts Whitehead's approach in rejecting static realms of clock readings, or a spatialization of time, by emphasizing actual or possible measurements ultimately based on experience (Fowler, 1975). "Static" time may be a better term than "Einstein" time since Einstein's own work involved elements of a process concept of time as indicated, for example, by his comments on the irreversibility of signal propagation (e.g., Einstein, 1949, 687).

For interpretations of the general theory of relativity, recent studies suggest that the coupling between the gravitational field and its sources should not be considered as uniquely supporting a geometric interpretation (Angel, 1981). Our failure to demonstrate enhanced coherence by assigning ontological priority to the gravitational field and its associated geometry or to its sources may indicate that they are complementary, if not coextensive, aspects of the web of events from which they emerge. A similar relationship occurs in quantum theory with the wave-particle "dualism," which is not a true dualism because it refers to emergent aspects of underlying quantum processes. Stapp provides an analogous explanation for the nature of physical or actual time, namely, that it arises from or is coextensive with more fundamental "sequences of actual becomings."

Stapp (1979) has stated that "the essential change wrought by relativity theory was precisely the rejection of the view of the world as a system developing in time, in favor of the overall spacetime view, in which one deals directly with the relationships between spacetime events." Process time is implicit in this overall space-time view because of the primary role of such relationships between events combined with the close yet asymmetric link between spatial and temporal intervals specified by the Lorentz relation. In relativity theory, time is not generally separable from spatial intervals and other possible relationships among events, which means that Einstein time is an abstraction from process time. Stapp states that Einstein time is the "time part of the space-time continuum of contemporary physics." This may be a useful definition, but it is misleading to suppose that Einstein time, so defined, involves a denial of process. If such a denial is presumed, then a metaphysical claim is being made which then introduces other problems (Čapek, 1971). Contemporary physical theory deals with the content of observations, each having information about various temporal and spatial intervals. But to what do those observations refer? The process of observation itself refers to events and processes.

In the initial draft of his paper for the Claremont conference, Stapp discussed two major philosophical shifts resulting from attempts to provide viable interpretations for quantum theory. These are (1) the shift from deterministic to probabilistic theory and (2) a change of ontological basis from the external world of geometric forms to the experiential world of observations. Quantum theory certainly made both these shifts difficult to avoid, although the many-worlds interpretation is one example of an effort made to retain a classical, multitudinist conception. However, developments in dissipative structures (Prigogine, 1980), classical systems exhibiting strong trajectory instability (Misra and Prigogine, 1983), as well as the quantum theory of measurement all suggest that the shift to probabilistic theory is linked to a fundamental time asymmetry in physical systems. This conclusion is not clearly independent of the formalism, although alternatives that attempt to deny an "open" future tend to be *ad hoc* and not coherent with human experience or with the overall framework of modern physical theory.

Stapp (1971) has previously made a distinction between "extrinsic" descriptions, which involve "effects of the system on its environment," and "intrinsic" descriptions, which consider the system as a self-contained entity. In quantum theory, the extrinsic description involves individual localized results (measurement preparation and descriptions of possible subsequent measurement), but these include no causal statements. The intrinsic description is causal but includes only predictions of probabilities (based on the products of density functions associated with the prepared states) rather than predictions of individual results. Bell's theorem shows that "demands of causality, locality and individuality cannot be simultaneously maintained in the description of nature." This dilemma is circumvented in quantum theory "by the use of two complementary descriptions, one of which omits the causality requirement, and the other of which omits the individuality requirement" (Stapp, 1971).

These complementary descriptions have a striking counterpart in Whitehead's philosophy. In part 4 of *Process and Reality,* "The Theory of Extension," Whitehead (1978, 433f.) discusses the coordinate and genetic divisions of the "satisfaction of an actual entity." In the genetic mode, prehensions (within which causality is subsumed) are exhibited in their genetic relationship and the actual entity is seen as a process. The genetic passage involves tendencies that may be associated with "sums over all spacetime paths" because the genetic mode is not in physical time, which arises in the coordinate analysis. Coordinate divisibility involves potential divisions so that the concepts of "bounded regions in spacetime" and the spacetime continuum apply to this coordinate mode. The prehensive unification that constitutes an actual entity thus requires both intrinsic, causal descriptions and extrinsic, localized descriptions. The counterpart forms of analysis are the genetic and coordinate modes.

Stapp's work has provided many provocative suggestions of how Whitehead's philosophy may provide a coherent philosophical framework for aiding the interpretation of the most recent results and implications of relativity and quantum theory. This critique of Stapp's paper is intended to clarify the relation of his very original work to that of Whitehead and the process-relational perspective.

Acknowledgment

In preparing this critique, I have benefited from several helpful suggestions and comments from the following colleagues at the University of Iowa: Professor William Klink and Dr. Crockett Grabbe of the Department of Physics and Astronomy, and Professors Evan Fales and Phyllis Rooney of the Department of Philosophy.

References

Angel, R. 1981. *Relativity: The Theory and Its Implications.* Elmsford, NY: Pergamon Press.
Bjorken, J. D., and Drell, S. D. 1965. *Relativistic Quantum Fields.* New York: McGraw-Hill.
Čapek, M. 1971. *Bergson and Modern Physics.* Dordrecht, Holland: D. Reidel.
———. 1975. "Relativity and the Status of Becoming," *Foundations of Physics* 5:607–17.
———. 1984. "Particles or Events?" In *Physical Sciences and History of Physics,* ed. R. S. Cohen and M. W. Wartofsky. Dordrecht, Holland: D. Reidel.
D'Espagnat, B. 1973. "Quantum Logic and Non-separability." In *The Physicist's Conception of Nature,* ed. J. Mehra. Dordrecht, Holland: D. Reidel.
———. 1983. *In Search of Reality.* New York: Springer-Verlag.
Einstein, A. 1922 [1956]. *The Meaning of Relativity.* Princeton: Princeton University Press.
———. 1949. "Reply to Criticisms." In *Albert Einstein: Philosopher-Scientist,* vol. 2, ed. P. Schilpp. New York: Harper and Brothers.
Finkelstein, D. 1973. "A Process Conception of Nature." In *The Physicist's Conception of Nature,* ed. J. Mehra. Dordrecht, Holland: Reidel.
Fowler, D. 1975. "Whitehead's Theory of Relativity." *Process Studies* 5, 3 (Fall):159–74.
Graves, J. 1971. *The Conceptual Foundations of Contemporary Relativity Theory.* Cambridge: MIT Press.
Grünbaum, A. 1967. *Modern Science and Zeno's Paradox.* Middletown, CT: Wesleyan University Press.
Misra, B., and Prigogine, I. 1983. "Time, Probability and Dynamics." In *Long-Time Prediction in Dynamics,* ed. C. Horton, Jr., L. Reicfil, and A. Szebehely. New York: John Wiley and Sons.
Prigogine, I. 1980. *From Being to Becoming.* San Francisco: W. H. Freeman.
Prigogine, I., and Stengers, I. 1984. *Order Out of Chaos: Man's New Dialogue with Nature.* Toronto: Bantam Books.
Stapp, H. 1971. "S-matrix Interpretation of Quantum Theory." *Physical Review D* 3:1303–20.

————. 1979. "Whiteheadian Approach to Quantum Theory and the Generalized Bell's Theorem." *Foundation of Physics* 9, 1–2:1–25.

————. 1982. "Mind, Matter, and Quantum Mechanics." *Foundations of Physics* 12, 4:363–99.

Von Weizsäcker, C. 1973. "Classical and Quantum Descriptions." In *The Physicist's Conception of Nature,* ed. J. Mehra. Dordrecht, Holland: D. Reidel.

Whitehead, A. N. 1978. *Process and Reality.* Corrected edition, ed. D. Griffin and D. Sherburne. New York: Free Press.

24. William B. Jones

PHYSICS AND METAPHYSICS:
HENRY STAPP ON TIME

Beginning at least with the publication of Copernicus's monumental *De Revolutionibus,* philosophers and scientists have struggled with the question of how to interpret the mathematical theories that modern physical science has put forward. Should they be viewed as accurate accounts of the nature of physical reality or as mere calculational devices, useful for predicting the future, for example, but hardly "true" in the literal sense of the word? In his preface to *De Revolutionibus,* Osiander opts for the latter alternative, but the verdict of history is that the metaphysical significance of such endeavors cannot be dismissed so lightly.

The problem becomes more acute in the case of Isaac Newton's incomparably successful mechanical theory. If Newton's theory is taken to be an accurate and complete account of the physical universe, then a number of the salient features of the commonsense world (viz., colors, sounds, smells, tastes) are denied objective status in the external, physical world and are relegated to the status of subjective human experiences. Hence, accepting Newton's theory as an accurate description of the physical universe leaves one with only two basic choices: dualism and materialism. That is, one can insist that colors, aromas, melodies, etc., are, indeed, real but admit that they belong to an internal or mental realm; or one can insist, for example, that colors *are nothing more* than electronic structures in the surfaces of things that tend to reflect relatively more of certain wavelengths of light, which, in turn, affect the retina of the human eye in particular ways, etc.—all of which can be described in purely physical terms.

The problem of interpreting Einstein's theory of relativity is even thornier than those encountered in dealing with previous theories. A fundamental feature of relativity theory is its denial that there is one, objective temporal ordering of all events. In some cases (pairs of events separated by "timelike" intervals), it may be possible to say absolutely that one event precedes another (and not vice versa), but in many cases (pairs of events separated by "spacelike" intervals) such absolute determinations of temporal priority are not possible. In such cases, temporal priority is not an objective feature of the events themselves but a function of the frame of reference employed in representing the spatial and temporal relationships obtaining among the events. The realization that temporal ordering may simply be a matter of how things are viewed has led some interpreters of relativity to wonder whether the same might not be true of the coming-into-being/passing-away aspect of things so fundamental to human experience. The supposition that this becoming/perishing quality of the commonsense world is really only a feature of human awareness (and not a characteristic of the objective, physical order) is reinforced by the basic mathematical structure of the theory of relativity. Relativity concerns itself with relationships in a four-dimensional space (the three dimensions of ordinary space and a fourth corresponding to time). This feature of the formalism of relativity theory has prompted a number of thinkers to complain that the theory involves "spatializing" time, banishing real becoming from physical reality. According to this interpretation, the four-dimensional space-time continuum of events just "is"; there is no objective, scientific basis for the commonsense notion that only the present is actual or real, that the past is gone, vanished, and that the future has yet to come into being.

Henry Stapp discusses this interpretation of relativity in an earlier draft of his paper, which was circulated as a basis for discussion at the Claremont conference on Physics and Time.[1] Although this matter is not addressed in the final version of Stapp's paper, which appears in this volume, it will be discussed here both as a way of giving the reader some sense of how the issues developed at the conference and as a way of coming to grips with the fundamental question posed at the beginning of this paper: given the mathematical theories of contemporary physical science, what can one conclude about the nature of (physical) reality? Or, how are the scientific and metaphysical enterprises related to each other? Efforts to provide an interpretation of the theory of relativity are particularly worthy of interest in that they call attention to some of the pitfalls awaiting those who undertake such interpretations.

In the earlier draft of his paper Stapp maintains that relativity entails neither three-dimensional becoming nor four-dimensional being. Rather, he argues, the theory of relativity is "mute" with respect to the metaphysical issue of whether and in what order things come into being. His reasons for taking this position will be discussed below, along with some related matters. At this point, it is useful to consider a very different reason for rejecting the thesis that relativity theory involves the spatialization of time, that it reduces physical reality to a four-dimensional space or continuum of events that just "is." This thesis should be rejected because it involves a fundamental error: the assumption that the "space" referred to when one speaks of the space inside a boxcar is "space" in the same sense as that referred to by a mathematician when he or she speaks of vector and function spaces of various dimensions. Depending upon the kind of space the mathematician is talking about, the geometrical relationships between points within the boxcar may embody or "model" relationships found in the abstract space of the mathematician, but it does not follow that the space he or she is talking about is exactly and fully "space"—just like that found in the boxcar. On the contrary, the mathematician is concerned only with the logical skeleton of such spaces and, of course, with many other such logical structures.

The basic variables of relativity have four components, corresponding to the four dimensions of the space-time continuum. Accordingly, they are called "four vectors." It should be clear from what has just been said that the four vectors of relativity theory are not vectors in a space just like ordinary three-dimensional space—except for having *four* dimensions. Thus, they are *not spatial entities*—at least not in *that sense* of spatial. And, of course, neither are they temporal in the usual sense. All that one can say, from a scientific point of view, is that these four-component variables are related in a definite way to the ordinary geometrical and temporal properties of things. The remarkable achievement of relativity theory is that it is able, by means of such an abstract mathematical formalism, to capture a vast array of relationships between these measurable geometrical and temporal quantities. The unsettling thing about the theory is the transformation properties of its four vectors; when these are subjected to a Lorentz transformation (as is in order when changing the frame of reference being used to one which is moving with respect to the original frame of reference), the geometrical and temporal aspects of things as we ordinarily conceive of them get *mixed up*. It is *this* feature of relativity theory which distinguishes it from classical Newtonian mechanics—*not* its formulation in terms of a four-dimensional spatiotemporal continuum of events. Indeed, classical (prerelativistic) mechanics can be, and was, formulated in terms of such a four-dimensional space. Consequently, the question of whether time really passes, or the four-dimensional continuum of events just exists in some timeless sense, can be (and was) raised in the context of prerelativistic mechanics.

Thus, one wonders why so many writers have viewed relativity theory as "spatializing" time. Perhaps it is because mathematicians use the expression *four-dimensional space* to refer to such systems; perhaps it is because the four vectors have three spatial components but only one temporal component; perhaps it is because everyone is so accustomed to seeing the time axis *represented spatially* in diagrams. In any case, if one is disturbed by the way relativity mixes up space and time, one might as well complain of the temporalization of space as complain of the spatialization of time. At least one ought to be prepared to grant equal time (or space!) to those who wish to make such complaints.

How is relativity theory to be interpreted? One could follow Osiander's lead and dismiss it as nothing more than a useful calculational scheme, but his example hardly inspires confidence in that approach at a time when astronauts confront directly things Copernicus could only guess at. On the other hand, if one accepts the view that the true character of physical reality is captured by the theory of relativity, then one must be prepared to deal with something that is neither spatial nor temporal but more fundamental than either spatiality or temporality. This conclusion is certainly not welcome news to philosophers who are committed to the view that coming-into-being/passing-away is a fundamental feature of reality. If temporality itself has only derivative status, it is hard to see how becoming/perishing, which would seem to presuppose temporality, could escape the same fate.

Anyone who is sympathetic to the idea that coming-into-being/passing-away should have a place in the basic scheme of things will be interested in the suggestion Stapp puts forward in his paper. Stapp submits that the time that is measured by clocks and that enters into the equations of physics, including those of relativity, is not the only kind of time. In addition to this kind of time, which he calls "Einstein time," there is, he postulates, a second kind of time, "process time." Most fundamental to Stapp's concept of process time is the idea of "becoming"—coming into existence, becoming actual rather than just potential, or becoming fixed and settled rather than merely confined to a range or set of possibilities. The "becoming" referred to here is not an observable process, however. If it were, it could be compared to other temporal processes (i.e., it could be "clocked") and it would be considered Einstein time.

The kind of "becoming" Stapp has in mind is perhaps best conveyed by considering the following question. Did all the events in the four-dimensional space-time continuum come into existence at once, or did they come into existence in some serial order (individually or in groups)? An alternative, more picturesque formulation of this question results if one thinks of each event as a dot in a four-dimensional Seurat-like painting and then asks whether this four-dimensional neo-impressionist canvas came into existence all at once or whether the dots appeared in some serial order. (Lamentably, this delightful, figurative way of putting the matter was deleted from the revised text of Stapp's paper.)

Of course, this question cannot be answered empirically. All measurable temporal relations are embodied in the "geometrical" structure of the four-dimensional space-time continuum. "Then," someone may protest, "the question is nonsense—rather like asking whether photons are the same color all over, as opposed to being striped or polka dot." And, of course, it *is* nonsense, *unless* there is a second kind of time, totally different from ordinary, measurable time, a fifth dimension, which is precisely what Stapp is suggesting. He is putting forward a metaphysical thesis, not an empirical or scientific one. Accordingly, reasons for accepting it will have to come from outside science. Thinkers who are concerned that time and real becoming have a place in the basic scheme of things will have to decide whether and to what extent "process time," as defined by Stapp, will serve their purposes.

A promising alternative approach both to the general question of what is to be concluded about physical reality, given contemporary physical theory, and to the specific question of whether (or how) real becoming fits into the basic scheme of things is to take quantum mechanics, instead of the theory of relativity, as one's point of departure. After all, there is no *a priori* necessity that attempts to frame a fundamental view of physical reality should take classical mechanics, even in its updated relativistic form, as its starting point. Its unparalleled success notwithstanding, Newton's mechanics was developed to deal with a very special class of physical systems, those which are isolable and reversible. But, as Prigogine has stressed many times, there are other kinds of physical systems for which successful theories have also been developed. Why not develop an account of physical reality based on one of them? It is this approach, utilizing quantum mechanics, that Stapp emphasizes in the final version of his paper.

At the most basic level, a prequantum-mechanical description of a physical system involves specifying the masses, positions, and velocities of the physical objects (particles) that make up the system and the forces acting upon them. From these and Newton's or Einstein's dynamical principles, the future (and past) states of the system can be calculated. (Of course, there are other, more formal and abstract formulations of classical mechanics, such as those worked out by Lagrange and Hamilton, but hardly anyone would have thought, prior to the development of quantum mechanics, that there was any direct correspondence between the basic elements of these abstract theories and physical reality.) A quantum-mechanical description of a physical system differs radically from its classical counterpart. In the simplest cases, such a description involves specifying the quantum state of the system. Each quantum state is labeled by a set of quantum numbers, which, when specified, are sufficient to identify uniquely the quantum state of the system. Quantum states usually reflect the symmetries and stability conditions of the physical system but do not correspond in any simple way to the properties of objects in the commonsense world. Given a quantum mechanical description of a physical system, one is typically able to calculate only the probability that things in the commonsense world will turn out a certain way.

A salient feature of quantum mechanics, especially important to anyone setting out to develop a view of physical reality based on it, is that there are a number of distinct formulations of the theory. These differ from each other markedly in their formal makeup but are generally experimentally indistinguishable; i.e., they have the same experimental implications. Perhaps best known in this regard are Heisenberg's matrix mechanics and Schrödinger's wave mechanics. Since they can be shown to be experimentally equivalent, the physical scientist is free to use whichever form is most convenient for his or her purposes, but the person attempting to work out a coherent view of physical reality is left wondering, for example, whether the wave function in Schrödinger's formulation, which has no counterpart in Heisenberg's matrix mechanics, corresponds to anything in physical reality. A conservative approach would be to attribute to physical reality only those characteristics which seem to be required by both (or all) formulations. Only very modest conclusions are likely to be obtained in this way, however. Clearly, more ambitious metaphysical enterprises will have to invoke criteria for acceptance from outside physical science. This point should be kept in mind when considering Stapp's specific suggestions toward a conception of physical reality that embodies the structure of quantum mechanics and provides a place for real becoming.

In developing his scheme, Stapp twice avails himself of the choices between alternative formulations afforded by quantum mechanics. The formalism of the theory can be developed in (four-dimensional) momentum space instead of (four-dimensional) space-time. Then the wave functions describing the quantum states of systems are functions of four-component, momentum-energy variables rather than of four-component, space-time variables. By choosing the momentum-energy formulation, Stapp eliminates any explicit reference to space-time and hence any need to address (at least initially) problems associated with it. To the physicist, momentum space is a postulated, abstract mathematical space (see the above discussion of the four-dimensional space of relativity theory), the acceptability of which is to be determined by the overall success of the entire theoretical structure of which it is a part. Stapp's metaphysical proposal is that it should be regarded as ontologically basic.

In a second choice of specific formulations of quantum theory, which is emphasized in the revised version of his paper, Stapp ties his metaphysical enterprise to an even more specific, and relatively little used, quantum-mechanical formalism: the S-matrix. The S-matrix is an array of mathematical functions expressing the connection between the various possible combinations of initial and final quantum states of at least certain kinds of physical systems. In the simplest case, the system might be a pair of particles, initially too far apart to affect each other but passing close enough to interact before receding out of range again. An S-matrix treatment of such a process does *not,* however, involve a description of what takes place *during* the interaction. It deals with only the states of the system before and after the interaction (specifically, its states in the infinitely remote past and infinitely remote future) and, of course, the mathematical connection between them. There are a number of (mathematical) forms such a treatment might take, but the S-matrix approach makes use exclusively of the momentum-space representation. The quantum states in terms of which the initial and final states of the system are described are states of definite momentum, i.e., states characterized by a momentum of specific magnitude and direction. States of this kind leave the spatial characteristics (e.g., location) of the system completely unspecified (as the uncertainty relation dictates). Furthermore, since only the asymptotic behavior of the system is considered in the S-matrix account, time does not enter the picture either. Nor is there any explicit dependence on energy, the dual variable of time.

In simplified terms, what the S-matrix does, in the case of a specific interaction, is to codify the relative likelihood that a system in a given initial (momentum) state will end up in each of a whole set of possible final (momentum) states. An actual physical system, suited to such a momentum-variable, S-matrix treatment (e.g., one particle approaching another), would start off in an initial state characterized by some definite momentum. Each element of the S-matrix appropriate to this system would express the probability that it will end up in a particular final state, also characterized by some definite momentum. The actual physical system would, of course, end up in *one* of these final states. (Eventually, the particle(s) would emerge from the region of interaction at some speed and headed in some direction or other.)

The crucial point for Stapp's purposes is that such processes involve a transition *from* a state of affairs in which there are a number of *possibilities* (possible final states of the system) *to* a state of affairs in which one of these possibilities has been *actualized*—to the exclusion of the others. Such transitions from a plurality of potentialities to a single actuality provide the basis for the sense of before-and-after that lies at the heart of Stapp's concept of "process time"; becoming actual, in this sense, is his fundamental metaphysical category. In Stapp's view, reality consists of a series of transitions of this type. Of course, if such instances of things "becoming fixed and settled," as Stapp phrases it, have fundamental ontological status, then so must the possibilities from which the actualities arise, as well as the "tendencies" in the nature of things that make it more likely that some of them will be realized rather than others. Thus, potentialities or possibilities must be taken very seriously if one follows Stapp's lead in this matter, just as is the case in his previous writings.[2] The Whiteheadian character of Stapp's view will surely be evident to everyone who is familiar with Whitehead's thought.

The tendencies in nature whereby some things are more likely to happen than others are the basis not only for all order in the universe, but also for time's having a direction and thus for there really being such a thing as time at all. For there to be an objective temporal ordering, it is not enough that all events have a polarity. They must also have an order. Physical reality, according to Stapp, "consist[s] of a well-ordered sequence of actual events." His S-matrix formalism incorporating the electromagnetic interaction provides for such an ordering but does not dictate what that ordering shall be. The specific vehicle of the ordering process is provided by the elements of the S-matrix, which express the tendencies under discussion. The basis for the ordering process lies in the mode of calculation of these elements, which involves the entire space-time continuum of relativity theory (e.g., the time axis from minus infinity to plus infinity). As a result, even states of affairs separated by a spacelike interval from the event being treated can make a difference in the S-matrix elements associated with the event and thus in its probable nature. Such effects, however, are limited to these tendencies or probability elements. The actual physical effects of an event, in particular the classical electromagnetic field (which is the only type of field explicitly treated by Stapp) associated with the event, are confined to a finite space-time region. More precisely, "[the occurrence of] event number n is represented within physical theory by a restriction on the set of classical fields allowed in the bounded space-time region R_n." That is, the classical fields inside the region R_n are restricted to some subset of the set of all conceivable classical fields in the region R_n. "This restriction induces, through the quantum formalism, changes in the tendencies for the next event. These changes in tendencies are manifested over all of space-time—i.e., even if the region R_{n+1} is spacelike situated relative to R_n." The preceding statement is as near as Stapp ever comes to enunciating an explicit temporal ordering principle. Unfortunately, it leaves some important questions unanswered.

Is the time sequence in which event number $n + 1$ is "next" after event number n process time or Einstein time? If the former, as Stapp surely intends, how does this ordering relate to that in Einstein time used in calculating the elements of the S-matrix? One possible answer is that they are totally independent. This answer would bring the present account into line with the earlier one formulated in the context of the theory of relativity, but it would imply that reversing the temporal priority of two events has no experimental implications. Such a con- clusion might be acceptable in a universe sufficiently rarified that every physical event is too remote from any other to be causally affected by it, i.e., a universe in which every physical event is separated from every other by a spacelike interval. Then one would be free to postulate an order—any order—in which the events supposedly came into being, in which the various transitions from a set of possibilities to an actuality took place. Such an ordering would not be empirically testable. The question of what significance it might have will not be addressed here.

But in a universe in which physical events can affect one another, the question of the order in which possibilities are actualized makes a difference. And this principle is just as valid in the context of a fundamentally statistical theory as it is in the context of a classical, deterministic one, as is evidenced by such atomic, nuclear, and solid- state phenomena as stimulated emission, cascade effects, and hole propagation. Furthermore, it is not hard to think of quantum-mechan- ical systems that, at one point in time, have the potential to go in two very different directions and that develop very differently if one of the two possibilities is realized rather than the other. One cannot be certain from Stapp's short treatment how such processes would be handled in a formalism such as his. As suggested earlier, approaches like the S- matrix have been employed heretofore almost exclusively in dealing with systems that can be usefully treated by considering only their asymptotic states, e.g., scattering phenomena of various sorts. In any case, if the transition from a number of possibilities to a definite actuality is to be taken as the fundamental physical reality and as the ultimate indicator of the direction of ("process") time, then systems that exhibit the kind of branching behavior mentioned just above must be regarded as exhibiting process-time ordering, an ordering in agreement with Einstein-time ordering.

This example, together with the others, suggests an answer to the question of how process time and Einstein time are to be related. Adhering steadfastly to the quantum-mechanical or "process" conception of time as fundamental, and assuming that there are not physical phenomena that cannot in principle be given a quantum-mechanical treatment, one might decide that in cases in which temporal ordering is empirically testable (in effect, in the case of pairs of events separated by timelike intervals) Einstein temporal ordering and process temporal ordering *coincide,* and that in cases in which temporal ordering cannot be established empirically (in effect, in the case of pairs of events separated by spacelike intervals) one is free to postulate, as part of an overall (S-matrix) quantum-mechanical account of reality, a definite temporal ordering of events, a definite order in which there have been transitions from a state of several possibilities to a state of one actuality.

This proposed account of time has the distinct advantage (over the one originally suggested by Stapp in the context of the theory of relativity) that it does not involve the metaphysical extravagance of postulating a new and experimentally inaccessible fifth dimension of physical reality. The postulated temporal ordering of events separated by spacelike intervals is, of course, a purely metaphysical thesis. It cannot be supported or refuted by empirical evidence. It must be judged not in isolation but as part of a larger picture, an encompassing metaphysical scheme. And the reasons for accepting (or rejecting) such encompassing systems must be drawn not just from empirical science but from a broad range of perspectives. It is, of course, characteristic of process philosophy that it endeavors to develop an account of reality integrating as broad a range as possible of human perspectives and experiences. The account of physical reality and time discussed above has the distinct advantage of being compatible with the most basic contemporary physical theories and, at the same time, of being ideally matched to the needs of process philosophers. Whether a successful quantum-mechanical conception of time can be developed that is not dependent upon the S-matrix formulation is a question that cannot be addressed in this short paper. But, at the very least, Stapp's efforts will help promote the idea, long championed by Prigogine, that classical mechanics/relativity theory is not necessarily the right starting place for an analysis of physical reality and time.

Notes

1. A copy of Stapp's earlier version, entitled "Time and Quantum Process," is available at the Center for Process Studies, 1325 N. College, Claremont, CA 91711.

2. W. B. Jones, "Bell's Theorem, H. P. Stapp, and Process Theism," *Process Studies* 7, 4 (Winter 1977):250–61.

25. David Bohm

As I have already mentioned in my own paper in this volume, the S-matrix approach of Chew and Stapp, in terms of graphs and Feynman diagrams, may be regarded as a particular way of interpreting the implicate order, as it arises in modern quantum-mechanical field theory. It has, however, the great advantage of providing a fairly precise and detailed formulation, which can moreover be related to current research on elementary particles and in cosmology.

Stapp has made an important step forward by showing that, according to the S-matrix theory, the electromagnetic field emitted by matter has a well-defined classical part (which is constituted of the "soft" photons). Understanding this opens the way for the development of a quantum theory in which observation and measurement do not play a fundamental role. Moreover, as I also said in my paper, from the point of view of the implicate order, one can say that Stapp has shown how the basically implicate structure of S-matrix theory contains a distinguished suborder associated with the classical part of the electromagnetic field, which defines what is essentially the explicate space-time order of special relativity. In this way, Stapp, in effect, offers a solution to the problem of how the implicate order can be taken as fundamental, without the need for starting with any assumed explicate order at all.

Stapp has further proposed a way of dealing with the irreversibility of physical process, by connecting his theory with the process philosophy of Whitehead. This he does by introducing a succession of actual events, e_n, which are located in relationship to regions, R_n, defined by the classical part of the electromagnetic process. This is a very interesting suggestion, which deserves to be explored in much more detail. In such exploration, a number of questions could usefully be considered.

First, it is clearly necessary to define the regions, R_n, in terms of physical process, as Stapp has indeed done. It is also necessary to characterize the events, e_n, in a similar way. How is this to be accomplished? Is there perhaps a "classical part" to the movement of particles (fermions), as well as to that of fields (bosons)?

More generally, it is not yet clear how events are to be defined at all. Perhaps defining events will require that the S-matrix theory be extended to a complex system, to include not only, as Stapp has suggested, large molecules, but also large-scale systems. Such systems could furnish suitable experimental contexts, in which the significance of events could be more fully determined. In the appropriate contexts it would be possible, from the laws of the quantum theory itself, to arrive at Bohr's approach, in which the form of the experimental conditions and the content of the experimental results (e.g., particular events) would constitute a single whole. We could then do justice to the extreme context-dependence of the meanings of events corresponding to measurements at a quantum level of accuracy (e.g., of spin), while we could at the same time recover the context-independence of the behavior of matter in the classical limit.

Stapp has suggested that his theory can be extended to include soft "gravitons," which would provide a classical part of the gravitational field. Extending the theory in this way would be particularly important, since the physical meaning of the space-time coordinates depends crucially on this gravitational field. Indeed, I feel that the extension of Stapp's theory to include gravitation could lead to new insights into the deeper relationships between the implicate order of the S-matrix (which would now include gravitation) and the explicate order of space-time. For after all, the electromagnetic field equations (both classical and quantum) can have their usual meanings only when there is such a classical part to the gravitational field. Therefore, the determination of space-time through the soft electromagnetic photons depends ultimately on the deeper ground of a basically implicate gravitational field, which has a special and distinguished explicate part.

Finally, there seems to be an unsatisfactory feature of the theory in that the quantum-mechanical tendencies (represented mathematically by the probabilities) depend on events in the indefinite future of a given event. However, as Stapp suggests, this dependence may, under ordinary circumstances, be too small to be important and may perhaps be further avoided by the use of an extension of Prigogine's approach of making an appropriate selection of bounding conditions. Nevertheless, some conceptual clarification is needed here. For example, are we returning to a modified version of Einstein's "block universe," in the sense that a specification of events in the whole of space-time is now needed for expressing the laws that apply to each event? If so, then it is not clear that this view will be entirely compatible with Whitehead's process philosophy. Perhaps the resolution of such difficulties depends on developing the theory further, so that it can provide a further answer to the question, "What *is* an event?"

26. Peter Miller

On "Becoming" as a Fifth Dimension

In *Process and Reality,* Whitehead distinguishes the genetic and co-ordinate analyses of actual occasions. A genetic analysis distinguishes and relates the various stages of the process of concrescence or creative becoming, whereas a coordinate analysis views the completed actual occasion as a spatiotemporal quantum occupying a regional standpoint within the extensive continuum. The two are distinguished because the process of concrescence does not proceed from earlier to later portions of the spatiotemporal quantum but presupposes and incorporates the whole quantum in all its phases from initial data to completed satisfaction. Furthermore, the coordinate analysis admits the formal properties of a mathematical continuum, including infinite subdivisibility, but the stages of genesis or becoming do not admit this type of mathematical analysis.

The distinction that Whitehead draws has led Patrick Hurley and Henry Stapp at this conference to identify "becoming" as a kind of "fifth dimension," i.e., as forming a sequence of before and after ("process time") that is utterly distinct from the before and after of physical time. Stapp's interpretation of quantum theory posits a single sequence of events in a well-ordered process time or order of becoming. He thus abandons the multiple independent sequences of becoming in a contemporary world whose problems Hurley faced. This single sequence of events, however, is determinative of a relativistic four-dimensional space-time continuum in nonlocal ways. That is, *each* process event narrows or restricts the tendencies of *all* space-time without regard to spatiotemporal contiguity and from minus to plus infinity in physical time, while and as it fully determines the actual properties of a particular bounded region of space-time. Although the revised paper does not specify the order of actualization of the bounded spatiotemporal regions, Stapp's conference presentation allowed that this too need not observe a one-way flow in physical time. That is, if event B is later than event A in physical time, it might nevertheless be the case that in the order of becoming or process time, A was actual later than B. Stapp presents this view as "becoming in four dimensions" to distinguish it from Newtonian "becoming in three dimensions" and Einsteinian "being in four dimensions."

The idea of becoming as a fifth dimension may be a forceful way of underlining the distinction between a coordinate and genetic analysis of an actual occasion, but there are serious difficulties with the suggestion that must also be considered. First, if the other four dimensions exhibit mathematical continuity, why should not the fifth as well? This, I think, is a problem for Whitehead himself and not only for Hurley's and Stapp's interpretations. There are several possible replies. Stapp (in conversation) takes the view that becoming satisfies a different mathematical model in that each actual occasion comes to be in a finite number of stages, each of which has a well-defined next stage, and that the larger series of occasions shares a similar order. Hurley, on the other hand, analogizes Whitehead's becoming to Bergson's *durée*. Bergson thought that any type of mathematical analysis of *durée* into discrete steps would falsify the data and so its unique character must be grasped intuitively.

Another problem of the "fifth dimension" or "process time" views is that they go too far in dissociating the order of becoming from the temporal order. Whitehead's distinction between the two applied only within the space of a single concrescence. In the larger picture, the temporal order of the extensive continuum is identical with the order of becoming. Although some of the topological and metrical specifics may be contingent features of the universe captured in some version of relativity theory, the extensive continuum *per se* just is the potential of each actual occasion to be objectified and prehended by others, and this, surely, reflects an order of becoming. Each actual occasion as it becomes occupies a standpoint relative to others that have already become (its past actual world) as well as relative to others that might become or might be becoming.

Since the past actual world of both process *and* physical time is decided, settled, and has perished, it cannot admit of further determinations by subsequent events. Stapp's proposals, then, are not just a minor adjustment to the theory but a radical break that creates a new problem regarding the metaphysical status of the extensive continuum (since presumably it can no longer have the status of defining the potential order of becoming that Whitehead assigned to it). A less radical departure would be to place restrictions on the formalisms of quantum theory that permit the determination of nonlocal potentials by actual occasions but only in not yet fully determinate regions of space-time, i.e., not in the past. I do not know how this possibility would fare in physical theory but it would seem to create fewer metaphysical difficulties for process philosophy.

27. Ilya Prigogine

A Short Comment on Henry Stapp's
Contribution

I very much appreciated Stapp's contribution, based on S-matrix theory. However, S-matrix theory, as usually understood, corresponds to a unitary transformation and to a symmetrical treatment of past and future. I am therefore somewhat at a loss to understand the difference between what Stapp calls "Einstein time" and "process time" on the basis of S-matrix theory alone.

Stapp comments that the physical origin of the symmetry-breaking mechanism remains obscure.[1] I would agree with this remark. Why our world contains irreversibility and life is indeed a question for which I have no answer.[2]

My purpose is much more modest. It is to start with the existence of irreversibility as a phenomenological fact and to see how to include it in a consistent formalism. To use an analogy, the reason why gravitational and inertial mass should coincide is far from clear. So is the fundamental meaning of Planck's constant. In many situations, progress in physics has been associated not with the elucidation of problems involved in the foundations, but in the incorporation of a given empirical situation into the framework of a general formalism.

Coming back to the S-matrix formalism, I should like to mention that in recent papers we have studied potential scattering from our point of view.[3] However, this is too technical to be developed here. Let me only mention that as a result of the nonunitary transformation, we obtain a representation in which the scattering cross-section plays the role of a generator of motion. This corresponds to a special case of the transition from groups to semigroups, which I have emphasized above.

Notes

1. However, there are situations where the microscopic meaning of irreversibility is clear. For unstable classical dynamical systems, the thermodynamical time is oriented in the direction of the expanding manifold.
2. Let me refer again to the paper by Géhéniau and myself, to appear in *PNAS,* in which we try to show that the direction of time is simply related to

the cosmological process (more precisely, to the dissymmetry between gravitational energy and matter).

3. C. George, F. Mayné, and I. Prigogine, "Scattering Theory in Superspace," *Advances in Chemical Physics,* 1985.

III Philosophical Overviews

28. Milič Čapek

THE UNREALITY AND INDETERMINACY OF THE FUTURE IN THE LIGHT OF CONTEMPORARY PHYSICS

Choosing the title for this paper caused me some difficulty. I was originally thinking about calling it "The Inescapability of Time in Physics," but this would have sounded rather truistic; for which physicist or philosopher of science openly and explicitly denies the reality of time? Even those who assimilate time to a fourth dimension analogous to the three dimensions of space emphatically and indignantly insist that they deny only *becoming* without denying time. Obviously, the word *time* has very different connotations in different minds. This is why I chose a different title—a title that would suggest more specifically two basic questions concerning the nature of time in physics that have received different answers by different thinkers. The first question is: does the reality of time imply the unreality of the future? The second question: if we admit the unreality of the future, do we have the right to regard it as being "open" in the sense of being undetermined, or is it still possible that such a nonexisting future is completely and un-ambiguously predetermined? It is clear that we can in principle obtain four different answers:

1. The future is determined and real,
2. the future is determined, but not yet real,
3. the future is undetermined, yet real, and
4. the future is both undetermined and unreal.

The history of both philosophy and science shows that all four answers have been given in different periods and by different philosophers and/or scientists. The oddest answer seems to be, at least at first glance, the third, since its intrinsic inconsistency seems to be immediately obvious: how can the future be real without being also determined? Yet this apparent inconsistency is not obvious to some serious philosophers of science of today, even though there are only a few of them. Furthermore, we shall see that this rather odd view was upheld by some respectable theologians in the past; it will not be irrelevant to mention them, since the dividing line between theology and cosmology in the Middle Ages was rather dim, if it existed at all.

Before considering the concepts of time in classical physics, it will be useful to recall some features of the notion of time as it appeared to unsophisticated common sense prior to the emergence of modern science and philosophy; for it is important to remember Whitehead's words that classical science is merely a refined form of common sense. The feeling of passing time is the most pervasive feature of our experience, both sensory and introspective; there is no question that this feature, at least in human beings, is closely correlated with the feeling of the unreality of future events. This feeling is very probably absent in the lower species of animals, but it is certainly present at least in higher mammals for which motor activity requires an antici- pation of future situations. It is difficult not to admit that a dog chasing a rabbit must have a rudimentary notion of the future; the contrast between the present sensory image of the running rabbit in the visual field of the dog, and the merely anticipated image of its capture, is the basis of what in humankind becomes a more abstract difference between the present and the future. The distinction is due to the fact that even the most concrete representation of an anticipated situation does not have the full sensory intensity of the present perception; in other words, every future situation appears as *not real*—more accurately, as *not yet* real. This correlation of the consciousness of the future with actions and anticipations has already been stressed by Jean Maria Guyau in his study *La genèse de l'idée du temps,* published a year after Bergson's first book, *Time and Free Will,* although it was written before it. The same author correctly stressed that when we are acting or actively planning, we have the feeling of moving *toward the future,* whereas whenever we have the attitude of passive expectation, the future seems to be moving toward us. Both kinematic metaphors are thoroughly misleading and impede our understanding of time; but it is interesting that they in either case suggest that the future somehow exists *prior* to our perception of it. This originally vague and inarticulated feeling became, by the process of generalization and conceptualization, the source of the conviction that the future state of affairs exists, or rather *preexists,* despite our illusory feeling that events are coming into ex- istence in a gradual, successive way. A very large part of the Western theological, philosophical, and—until recently—scientific tradition was dominated by this conviction.

There is no place here to give a detailed description of the develop- ment of this view. Suffice it to say that it was present already in the mythological belief in prophecies, according to which what may be called "distance in time" is transparent to some privileged minds, as they are able to perceive the events deep in the future. Even today the prophecies of Nostradamus are taken seriously by many people and his books have been selling very well recently. The role of prophecies in Christianity as well as in other religions is well known. God's foreknowledge of the future has always been a part of Christian as well as Moslem dogma; the doctrine of predestination was a mere consequence of it. Proclaimed already by St. Paul, probably under the influence of Stoic determinism, accepted by St. Augustine, St. Thomas, and the Protestant reformers, it is still being accepted—and not only by Calvinists; in truth, it is accepted by all but process theologians. Even among them there are some who remain nostalgic for the allegedly noble notion of divine omniscience and try to combine Whitehead with Boethius.

Interestingly, this notion was retained even when philosophy emancipated itself from theology. When the concept of Supreme Being became secularized in the pantheism of Bruno and Spinoza, its time-transcending character was fully preserved. This continued even in modern post-Kantian idealism and in English neo-Hegelianism. This is what Charles Hartshorne has called the "Philonian tradition" in theology and philosophy; I think it would be historically more correct to speak of the "Eleatic" or "Parmenidian" tradition, since this notion can be traced back to the ancient doctrine of Parmenides, who was the first to insist upon the immutability and timelessness of "true reality." Since Parmenides the obsession with "the timelessness of reality" has persisted through the centuries. Although only very few followed Parmenides in his radical denial of time and plurality, nearly every philosopher, beginning with Plato, has assumed that "the true reality" is beyond time and that change—which was most frequently and onesidedly identified with "corruption" and "decay"—belongs to the "imperfect" world of sensory perception or, as Kant later called it, the "phenomenal" world. In other words, this tradition meant that, from the list of alternatives mentioned above, the first one was adopted: *"the future is determined and real."*

Aristotle, it is interesting to note, belongs to this group of thinkers only in part. True, his Prime Mover has all the Eleatic and Platonic attributes: it is beyond time. But he also explicitly stated that the future is *both unreal and undetermined,* at least in the sublunar world. He, the first systematizer of classical logic, did not hesitate to deny the applicability of the law of the excluded middle to propositions referring to future situations. This he illustrated by his famous example: the proposition "There will be a sea fight tomorrow" is neither true nor false today; its truth is now uncertain—not only because of human ignorance, but because *the future itself is objectively ambiguous.* Aristotle thus came very close to the modern three-value logic, as well as to the objectivistic view of probability in present physics; he is, at least in his *De Interpretatione,* probably the first genuine process philosopher— far more than Heraclitus, whose dynamic philosophy is marred by the notion of eternal recurrence, by which becoming is converted (implicitly at least) into being. Mircea Eliade rightly observed that this doctrine is "a manifestation of ontology uncontaminated by time and becoming." Cyclical time, in which the future has already been past and the past will be future an infinite number of times, is no time at all.

One may be tempted to conjecture that Aristotle's option for the open future was due to his sense for concrete experience, since he was not only a metaphysician but a physicist as well. But this was not true in this case; his rejection of determinism was motivated by moral considerations only. He regarded the concept of a strictly predetermined future incompatible with freedom and ethical responsibility. More than two thousand years later the same motives led William James to reject the deterministic "bloc universe." Despite their enormous distance in time, James and Aristotle agreed in their conclusion: *the future is undetermined and therefore unreal.*

Unlike theologians and philosophers, the great majority of scientists of all ages have accepted the reality of time and its corollary—the unreality of the future. This was true even when they accepted strict determinism, whether naturalistic or theological. But it is rather significant that ancient atomists as well as the medieval cosmologists preferred to accept *the relational theory of time.* In other words, time for them was inseparable from, and dependent on, concrete events; it did not have any substantial reality. Epicurus called it *"accident of accidents,"* maintaining that it does not exist apart from atoms: fundamental reality belongs only to space and atoms. In Aristotelian cosmology, time was associated, sometimes even identified, with the uniform rotation of the last celestial sphere; it would be nothing without the celestial clock. These views apparently have a modern ring, as they seem to have a certain affinity with the post-Newtonian rejection of absolute time; but a closer analysis will disclose that the alleged similarity is rather superficial. Take, for instance, the basic premise of atomism: there is nothing in the world except matter and the space *in which it moves.* But please note the words I have italicized: it is clear that the existence of motion is *assumed,* and the concept of motion implies that of time. It is logically impossible to obtain from a bare concept of space and from a bare concept of matter the concept of time. It is true that materialists of all ages have insisted that there is no such thing as matter without motion and, consequently, without time; but this is a mere statement of fact, a mere empirical correlation, not a logical implication, as both Leibniz and the early Bertrand Russell pointed out.[1] The question is as follows: is it logically possible to build an atomistic, or, more generally, a kinetic-corpuscular model of nature on only two basic concepts—that of space and matter—without surreptitiously smuggling into it also the concept of motion and *ipso facto* the concept of time? Is it not true that it is impossible to think of matter and space *without their being in time?* In other words, as soon as we think of space—even of empty space—as *enduring,* we think of it as *enduring in time.* Even if we assume matter is completely stationary—which was certainly possible in classical physics at least (just remember the concept of motionless ether or Carl Neumann's "body alpha")—as soon as we speak of its permanence we must think of it as *persisting through time.*

This fundamental ontological importance of time was fully grasped by Newton and some of his predecessors, such as Isaac Barrow and Pierre Gassendi. Time thus acquired the same ontological rank as space as they both became *sensoria Dei,* the physical embodiments of God's omnipresence and eternity. It is interesting how Newton's predecessor, Gassendi, and his follower, Samuel Clarke, both stressed the enduring, i.e., temporal, nature of Divine Being; they both criticized and even ridiculed the theological concept of "Eternal Now," which they dismissed as "a mere play of meaningless words" *("lusus merus non intellectorum verborum").* Gassendi also anticipated a more correct translation of the famous passage from *Exodus,* which should be read "I shall be what I shall be" instead of "I am what I am."[2] The temporalistic conception of the universe as well as modern process theology did have its predecessors in the seventeenth century.

But the philosophical impact of such views was rather limited. The concept of timeless reality was too well entrenched to yield to the opposite view; Newton's elder contemporary, Benedict Spinoza, was only half-Newtonian, as he assigned a fundamental ontological status to space, but not to time. God was for him *substantia extensa,* but not *substantia durans;* his eternity was clearly timeless. But the decisive reason for a continued degradation of time was the triumph of scientific determinism. With the coming of mathematical physics the deductive concept of causality prevailed; the causal relation was viewed as a relation of *logical implication.* In other words, every state of a mechanical system was regarded as being *logically contained* in its antecedent state, from which *all* its features can be deduced; the ancient scholastic expression *"causa aequat effectum"* acquired a precise meaning, which became a model for every causal explanation—not only in physics, but gradually in every science. Long before the mechanization of the world picture was completed, Spinoza applied it to the whole of nature in two famous propositions, the twenty-ninth and thirty-third of the first book of his *Ethics:* in nature there is no contingency since everything that happens *happens necessarily and cannot happen in any other way.* Unlike Descartes, Spinoza realized that in the mechanistic universe there is no place for human freedom nor for any of the spontaneity we usually associate with organic life. The epilogue to the first book of his *Ethics* reads like a prologue to Darwinian mechanistic philosophy; no place whatever is left for teleology or for any nonmechanical agency.

Furthermore, Spinoza was among the first to realize that *the deductive concept of causality virtually eliminates time.* He says that the world follows from the eternal substance of God in the same way as from the definition of a triangle follow the specific geometrical theorems about it (such as the theorem about the sum of its internal angles). But the words *it follows* should not be taken in a temporal sense; it is a logical, not a temporal, sequence. Spinoza says that "in eternity there is no *when,* nor *before,* nor *after" (cum in aeterno non detur quando, ante nec post);*[3] but he merely thereby expresses in metaphysical terminology the fact that the logical implication is by its own nature timeless. The conclusion does not follow from its premises *in time,* but in a tenseless way; more correctly, it is *contained* in, or *implied* by, the premises. The fact that we grasp the conclusion after its premises is irrelevant and is due to our limitations. It may be objected that Spinoza was a metaphysician and not a scientist; there is no question, however, that he was deeply interested in the incipient sciences, especially in mechanics and optics, as Richard McKeon established beyond doubt; furthermore, in the seventeenth century, the dividing line between scientists and philosophers was rather dim. After all, Spinoza restated what Galileo proclaimed before him: it is we who are condemned to think in time as we proceed in a step-by-step fashion from inference to inference, whereas the divine insight is timeless as it grasps the truth "without any temporal discourse" *(senza temporaneo discorso).*[4]

Galileo retained the duality of God and the world—the duality which Spinoza abolished. In either of them the expression *it follows (sequitur)* is stripped of its original temporal connotation, since it acquires the strictly logical meaning of tenseless logical implication. In other words, instead of saying that the effect follows from its cause, we should say that it is *contained* in it. This may be also called "the container theory of causality," which became gradually dominant—as various historians of science, in particular Emile Meyerson and Arthur Haas, have established in a convincing and documented way. This view of causality culminated in Laplace's frequently quoted passage, according to which the impersonal world order became a timeless implicative pattern in which the distinctions between the past, present, and future disappear. Spinoza as well as Parmenides would have been delighted. Note the conceptual evolution: Galileo still spoke of God; Spinoza provided a substitute, speaking of "God or Nature" *(Deus sive natura);* finally, Laplace, in his famous answer to Napoleon, dropped even the word. ("Sir, je n'ai pas besoin de cette hypothese.") But Laplace's "omniscient mind" was really nothing but a secularized and depersonalized version of the omniscient God of traditional theology, equally changeless, equally timeless. How much this mode of thinking is alive today is shown in Einstein's words, pronounced at the very end of his life when he heard about the death of his close friend: "the distinction between past, present and future is only an illusion, even if a stubborn one."

The deductive theory of causality was accepted despite David Hume's doubts about the possibility of establishing any necessary link between successive events. For the development of physics it was probably better that the eighteenth-century scientists ignored Hume's criticism of causality; it is difficult to imagine Lagrange and Laplace writing their treatises of analytical mechanics had they been paralyzed by Humean doubts about the deductive model of causality. Furthermore, in the light of repeated triumphs of determinism, in particular of its spectacularly verified predictions based on deduction, Hume's epistemological doubts looked like armchair irrelevancies, at least in the eyes of astronomers and physicists. There is no question that they all accepted the deductive model of causality more or less dogmatically until 1927; some of them, such as Einstein and Planck, even after that.

Another of Hume's prophetic insights has been generally unnoticed. It can be found in his *Treatise of Human Nature* (book 1, part 3, section 2), where he pointed out that a complete simultaneity of cause and effect would imply "the destruction of succession" and "the utter annihilation of time." He wrote that "if one cause were contemporary with its effect and this effect with its effect and so on, 'tis plain that there be no such thing as succession, and all objects must be coexistent." We have seen that neither Laplace nor Einstein hesitated to accept this conclusion whereas Hume had been too empirically minded to accept it. In this passage he implicitly anticipated Bergson's charge that classical determinism virtually eliminates succession, since it reduces time to a mere subjective incapacity to know everything at once. Bergson, of course, was not alone; he was preceded by a few courageous thinkers in the second half of the last century, such as Charles Renouvier, Émile Boutroux, Charles S. Peirce and, in particular, William James, who had been influenced by Renouvier. But this occurred more than a century after Hume. Hume himself failed to draw any consequences from his remark; he did not even use it in his criticism of the concept of causal necessity, although it would have been a very cogent argument. This had to wait for the philosophers just mentioned. William James rejected, it is true, the static "bloc universe" mainly on ethical grounds; but in his little-discussed essay "On Some Hegelisms" he stated his objection in a superbly lucid way:

Why, if one act of knowledge could from one point take in the total perspective, with all mere possibilities abolished, should there have been anything more than that act? Why duplicate it by the tedious unrolling, inch by inch, of the foredone reality? No answer seems possible.[5]

James's target here was clearly the timeless idealism of neo-Hegelianism, but his objection also applies to the strict determinism of naturalism; both modes of thought make time a mere "tedious unrolling of the foredone reality," utterly superfluous. Charles Hartshorne once spoke of "the secret alliance of theology and naturalism," since they both accept the most rigid kind of determinism. We can equally well speak of an alliance of timeless idealism and timeless materialism; Bradley and McTaggart eliminated time as ruthlessly as had Laplace. It is true that most rank-and-file scientists were too empirically minded to deny the reality of time; to them the future is both determined *and* unreal. They clearly failed to grasp what the deductive view of causality implies: if the effect is *implied* by its cause, it should be *simultaneous* with it, and any delay of its happening is incomprehensible, since a delay is incompatible with the very nature of implication.

Today classical determinism has ceased to be a dogma in physics and can hardly be used as an argument for the reality of the future. The titles of two recent books written by two outstanding scientists— Hermann Weyl's *The Open World* and Karl Popper's *The Open Universe*—are both eloquent and significant. But what about Einstein's view? And what about relativity theory, which supposedly remains deterministic and is thereby often regarded as a culmination of classical physics? In answering the first question, one must keep in mind that Einstein in his last years discernibly wavered in his commitment to determinism. This was almost certainly due to the prodding of Karl Popper and perhaps also to the delayed influence of Bergson, mediated by one essay written by Louis de Broglie before his reconversion to determinism.[6] As far as the alleged deterministic nature of relativity theory is concerned, it appears as such because of its *macroscopic* character. It is known that the atomicity of Planck's action, *h,* which accounts for the discontinuity and indeterminacy of microphysical processes, can be safely disregarded on the macroscopic and *a fortiori* megacosmic scale. But at the same time let us not forget that among the most spectacular verifications of relativity were those made in the region of huge velocities, i.e., in microcosms where the atomicity of action plays such a decisive role. A mere coincidence? What is certain is that action (the product of energy and time) is one of a few quantities of mechanics that remain invariant in relativity. As Eddington observed long ago: "Erg-seconds or action belong to Minkowski's world which is common to all observers, and so it is absolute." All other quantities prominent in prerelativity mechanics refer to the three-dimensional instantaneous sections of space-time, which are different for different observers; "they need to be multiplied by a duration to give them thickness before they can be put into the four-dimensional world."[7] The Russian physicist J. Frenkel even submitted the proof, which so far as I know has not been challenged, showing that the atomicity of action is implied by Lorentz transformation. Since the indivisibility of action implies the indeterminacy of microphysical processes, this would indicate that, contrary to some popular and frequently held claims, the alleged incompatibility of the relativity and quantum theories simply does not exist. Furthermore, the rise of the new quantum theories such as wave mechanics was inconceivable without relativity.

Yet the relativistic concept of space-time is still being used in the alleged proofs of the becomingless character of the physical reality as well as of the rigorous determination of the future. This is the famous "myth of frozen passage," which I have already analyzed several times; but as Paul Weiss wrote me once in a private letter, it looks like the myth will be with us forever. It is one of the most inveterate habits of our thought and, as mentioned above, can be traced to the very dawn of the Western tradition. Before restating briefly the criticism of the theory of the "mind-dependence of becoming," as it is called, it is important to show that there are two versions of it. One version regards the concept of timeless or becomingless reality as synonymous with the rigorous predetermination of the future. This is certainly self-consistent; it is impossible to ascribe any objective ambiguity to the future if it exists prior to our own awareness. This is the view that goes back to Laplace, Spinoza, and eventually to Parmenides. But there is another version of the becomingless view (held, for example, by A. Grünbaum, Donald Williams, J. J. Smart), which claims that an elimination of becoming from the physical world does *not* imply a complete determination of the future. How a completely preexisting future can retain any kind of ambiguity or indetermination is simply a mystery— but it is not for the first time in the history of ideas that such a self-contradictory view was upheld. Quite a number of theologians, including Aquinas, accepted a complete divine predestination alongside human freedom; accordingly, the future was supposed to be real, though undetermined. They usually added "real for God, not humanity"; this would hardly save the self-consistency of their view, since by the same token it would reduce the feeling of the passing of time—together with the "futurity of future," without which freedom is impossible—to mere human ignorance.

As far as the argument for subjectivity of becoming is concerned, it usually runs along the following lines: there is no objective "Now" in relativistic space-time; the so-called "Now-line" separating the past from the future is different for different observers and consequently there is no objective basis for the reality of becoming. It is true that there are no objective "Now" lines; yet it is not the whole truth, and this is precisely what makes the conclusion wrong. Even a superficial inspection of relativistic time-space clearly shows that the past is separated from the future *even more effectively* than in classical physics: the whole four-dimensional region called "Elsewhere" (which could equally well be called "Elsewhen") is interposed between the causal past and causal future. It is this region which is differently divided by different observers and in which the "Now" lines do not coincide. But no "Now" line of mine can ever cut my own causal past nor my own causal future. Furthermore, every "Now" line is nothing but a mere ideal fiction devoid of any concrete physical reality. The apparently mysterious character of Einstein's universe disappears when we realize that his denial of absolute "Now" is a denial of *instantaneous* actions and of *instantaneous* space; the latter is nothing but an arbitrary instantaneous cut in the four-dimensional becoming.

Moreover, although there is no absolute "Now," there is such a thing as an *absolute future*. This is quite obvious for my own "Here-Now," which I actually experience; I can never perceive my own future event, since there are no backward causal actions, no effects preceding their causes. But more than that: my own causal future cannot be *perceived by any other conceivable observer*. By "conceivable observer" we can mean only an observer located somewhere in the "Elsewhere" region.

By the very definition of "Elsewhere," no causal actions from my own "Now"—*a fortiori* from my own future—can reach the observer unless they move with a velocity greater than that of light, which is excluded by relativity. Conversely, *for the same reason,* no event from the causal future of any observer located in the "Elsewhere" region can reach my own "Here-Now." Consequently, future events are *intrinsically unobservable;* hence, to accept their existence runs against all accepted rules of scientific methodology.[8]

It is, of course, true that intrinsically unobservable entities can in principle be postulated, but only when they contribute to the understanding of reality. In this particular case, the very opposite is true; the theory of the "mind-dependence of becoming" only enormously complicates the total picture. It creates a kind of dualism even *worse* than that of Descartes, a realm of becomingless physical being existing concurrently with a realm of subjective becoming. Descartes's two substances, no matter how disparate, shared at least their being in time. This is not so in this case; there is absolutely no way to correlate two such completely different realms. This is the intrinsic difficulty inherent in the ancient myth of Parmenides: how can the reality that is by definition becomingless become—note the word *become*—an appearance that is unexplainably *diverse* and *changing?* No answer is given and no answer is possible, since the question itself was needlessly generated by the invention of useless and fictitious entities. Future events are neither real nor determined; conceptual analysis and all available empirical evidence both point to this conclusion.

Notes

1. Leibniz's letter to Thomasius, Sept. 26, 1668, in *Leibniz's Philosophischen Schriften,* ed. C. I. Gerhardt (Berlin, 1875), 1:9–11; B. Russell, "Is Position in Space and Time Absolute or Relative?" *Mind* 10 (1901):294.

2. Pierre Gassendi, *Syntagma philosophicum* (Lugdunum, 1658), 227. Samuel Clarke quotes Gassendi's view in *A Discourse Concerning the Being and Attributes of God* (London, 1732), 43.

3. Proposition 33, scholium 2.

4. *Opere,* VII, 129.

5. William James, *The Will to Believe and Other Essays in Popular Philosophy* (London: Dover, 1956), 271.

6. For Einstein's comment about Louis de Broglie's essay on Bergson, see M. Čapek, *Bergson and Modern Physics* (Dordrecht: Reidel, 1971), 299–300; also my introduction to the forthcoming translation of Emile Meyerson, *The Relativistic Deduction* (Dordrecht: Reidel, 1985), xliv.

7. A. S. Eddington, *The Nature of the Physical World* (Cambridge, 1933), 180.

8. Cf. my article "Relativity and the Status of Becoming," *Foundations of Physics* 5 (1975):607–16; also, "The Status of Future Events," in *Abstracts of the Seventh International Congress of Logic, Methodology of Science at Salzburg* (1983):21–23.

29.   Frederick Ferré

On the Ultimate Significance of Time
for Truth, Goodness, and the Sacred

Let us back away from the detail that has inevitably engaged our attention in the discussions that have occupied us in this book. What outlines of significance emerge when physicists and philosophers ponder the nature of time in the universe? I plan to offer brief personal reflections on three dimensions that must interpenetrate at the point of our interests. These three dimensions have to do with *thinking itself* (or methodology and epistemology), with the *valued itself* (or aesthetics and ethics), and with the *divine itself* (or religion and theology). Such great issues are obviously too large for anything but the broadest of brush strokes here, but occasionally there is need to risk attempting this kind of sketch, even when one's strokes are tremulous and smudged.

29.1. Time and Truth

All of us—scientists, philosophers, theologians—want to think truly about time. We also know that time affects what is taken to be true. Without theories, without concepts, without language, we are helpless to formulate our would-be truths; but all theories and concepts and languages are historically conditioned. The available evidence, on the one hand, is a function of our historical era; the available conceptual forms, on the other hand, through which we take account of whatever evidence we have, are also time-bound through and through.

Thinking about time forces us to acknowledge a "dilemma of formalisms" that afflicts all human thinking. On one horn of this dilemma we see that conceptual formalisms, i.e., words, symbols, diagrams, models, and the like, are essential to our thinking. Without them we would be imprisoned by the vague and inexplicit; with them we gain the power of generalization and discrimination. Formalisms are necessary conditions for theory, and since *true* is an adjective that we hope will increasingly apply to our theoretical assertions and descriptions, we urgently need the best formalisms we can obtain or construct. On the other horn of our dilemma, however, we realize that all our past formalisms have proven to some degree inadequate. The passage of time has revealed their limitations, sooner or later. Perhaps our currently favored theories are merely the ones whose limits have not yet been revealed to us. To what, then, can *true* apply without qualification?

There is a reason, in principle, for this second horn of our dilemma. Formalisms are, by their very nature, abstractions. If they did not stand away from and simplify the clutter of full immediate experience, they would not help our minds gain a hold on the essential, the recurrent, the underlying, or the surprising-in-context. But to the extent that our formalisms simplify they also leave out and ignore. For many purposes such ignoring may be exactly what is needed, but in this necessary discrepancy between the abstract formalism and its concrete referent lies a time bomb for those who forget the inevitable partiality of the conceptual materials from which theories are constructed.

This does not mean that there are no better and no worse formalisms. The fact that all fall short—and must fall short—of full inclusion of all the data is no ground for giving up the search for increasingly apt abstractions and representations. This is exactly what the story of all theoretical change recounts, whether in philosophy or in science. Presently this process of refinement or replacement is going on in physics and the philosophy of science concerning the proper formalisms to use in the representation of time itself.

Until recently in our era, the formalisms through which time had to be represented in our physical theories were drawn in large part from the mathematics of Newton and Leibniz and from geometrical axioms that have hardly changed since Euclid. In particular, the assumption of mathematical "denseness" in calculus, which was developed to deal with problems of dynamics in the seventeenth century—the assumption that between any two instants there can be postulated an infinite number of further instants (just as in Euclid's geometry one can postulate an infinite number of points between any two points, however close)—led to a conception of actual time as sharing these punctiform characteristics. This was a natural assumption, since the formalisms of familiar mathematical tools have for physicists been the primary conceptual means through which to think about and calculate time-dependent processes. Should the processes themselves, then, not be assumed to be isomorphic with the useful (indispensable) formalisms of mathematics?

This natural assumption, however, often overrides the firsthand acquaintance we have of time, which never presents itself in durationless instants but in a "flow." Time comes to experience in minimal whole durations rather than as a "densely ordered" continuum. Adolf Grünbaum, though a prominent voice, is just one among many who acknowledge the incompatibility between human temporal experience and standard mathematical formalisms but choose to follow formalism rather than to trust experience. The result is a conception of the universe on the model of the necessities of geometry. Albert Einstein himself so conceived the universe of space-time, in which all events have a place in a continuous four-dimensional plenum, existing tenselessly at various clock-times. The differences human experience finds between past, present, and future, on this conception, are merely "mind-dependent." In reality, the theory claims, if there were no minds to cast a subjective hue—worry over the future, satisfaction or remorse over the past—on events, there would be no such thing as past or future events. There would be only earlier or later events spread across the great cosmic continuum in which, on one dimension, events are distinguished not only by spatial but also by "timelike" separations.

My mention of Einstein's name in connection with this conception of time shows that it is a distinguished view. Indeed, it has in recent years been virtually the only scientifically thinkable theory. And yet it rules out, if it is accepted, fundamental human intuitions of the unfinished character of the future as contrasted with the settled character of the past and the decisive, event-actualizing character of the present. On this view, there is no reality exposed by such intuitions. The future and the past are wholly symmetrical in themselves (if not from our mind-dependent point of view), and there is no such thing for the universe as the present.

It would not be too strong, perhaps, to say that this reversible, becomingless conception of time is the standard view of modern science. We need not be surprised, therefore, at the sense of alienation that many have come to feel toward such science—many, that is, who are convinced of the importance of human intuitions into the decisiveness of the present and the openness of the future to various possibilities and into the objective nonreversibility of the present and the past.

This is what makes the work of certain physicists—like Prigogine, Bohm, and Stapp—so profoundly exciting and important. They are challenging at its root an orthodoxy on how to conceive of time. It is an orthodoxy that dates at least from Newton's invention of the calculus. These thinkers, however, are offering to replace the standard conceptual tools of modern physics and to pave the way, it might be said, for a post-modern physics that no longer forces the choice between vital human experience and the powerful clarities of theory. Prigogine's work with nonreversible processes, with what he calls the "physics of becoming," for example, gives ground for the affirmation of an objective difference between future and past for the physical universe, in keeping (at least in principle) with human intuitions into the significance of responsible agency.

A long road lies between these beginnings of post-modern physics and the full integration of science and the humanities. What is compatible in principle needs to be connected in detail by theorists who approach with the breadth of vision required by metaphysics or theology. There may be generations of work to be done to fulfill the vision of post-modern integration, just as there were generations of work required to fashion the mechanical world view out of the brilliant but overly abstract insights of the seventeenth century.

What recent developments have shown us, however, is that theologians and philosophers and poets were right to resist unlimited application of the formalisms of modern physics to the universe as a whole, even when this was attempted by distinguished and prestigious thinkers. The passage of time has allowed the formulation of new possibilities for the conception of time itself and thus of the universe. If we are concerned for truth, then, we must be ready not only to apply but also to challenge all formalisms. This dual role, indeed, is the function of philosophy. On the one hand, science, as it develops and changes over time, sets a context for responsible philosophical speculation that uses the best scientific abstractions of an era and pushes them as far as they can go; but on the other hand philosophy, as critic of abstractions, subjects science to continuing scrutiny in terms of the fuller data of experience as a whole.

One side of the quest for truth, then, requires seeking for understanding that can only be supplied through concepts and symbols and models—abstract formalisms that are fashioned painfully from the devotion of genius, the precious heritage of the ages. On this first side, the quest is for *coherence* among such abstractions, and successful theorists rejoice in the clarities that emerge from the deft manipulation of concepts illuminating our experience of the world. The other side of the quest for truth, however, demands attending to the complex and massive data that we receive from the world we want to understand. We must remember that all our formalisms, however successful, are partial renderings of this domain. On this second side, the call is for *adequacy* to the data of experience. It is appropriate that A. N. Whitehead once warned: "Seek simplicity, and mistrust it." In that duality the dynamics of truth-seeking are summarized.

This means, of course, that now as we rejoice in the more adequate formalisms of a nascent post-modern physics, we need to keep on critical guard against the reification even of these welcome abstractions. But this caution is built into the ultimate significance of time for the quest for truth. Even our favored thoughts about time are temporal—not necessarily wholly wrong as a result, but certainly in need of being surpassed in the long run that time provides for thought.

29.2. Time and Goodness

There are various kinds of goodness, and time is essential for them all. The one possible exception to this rule is the goodness of certain aesthetic achievements, like great paintings or sculpture and towering mathematical constructions, that seem in their sublimity to escape from—even to defy—the temporal flux. In one sense, of course, Plato's vision of the best as beyond change and decay rests here and needs to be incorporated in any adequate account. I shall return to the question in connection with the sacred.

But what of aesthetic goodness? Some aesthetic objects, like a poem or a symphony, are in themselves temporal. This is a large class, including dance, drama, film, the novel or short story, and all music. The very fabric that constitutes these creations is woven on the loom of time. Without time they could not be at all what they are.

The same, I think, is true for the aesthetic value even of the unchanging formal patterns in painting, sculpture, architecture, and mathematics. An essential ingredient in our experience of these achievements of noble stability is the necessary temporal background of our deepening appreciation of the relationships among juxtaposed elements. Not only did these achievements take time in the making—the building of the temple, the chipping at the stone, the working out of the proof—but also, and likewise, they require time for their aesthetic realization.

Whatever our views on aesthetic goodness, there can be no doubt that moral goodness requires time as its necessary condition. The building of a moral character takes a lifetime. Each significant moral decision that goes into the shaping of such a moral life requires time in two senses. First, a decision of this sort needs a duration of time for its deliberation. If it were an automatic reflex, it would not count as a decision and would not be subject to moral praise or blame. Alternatives need to be distinguished, consequences predicted, principles weighed. Even in a mature moral agent, this must take some time if a genuine decision is involved. Moral goodness, then, depends upon time for deliberation. Second, if deliberation is to be real and not a hoax, the moral agent must be in a position to choose between possible courses of action that are genuinely open. If there is only one possible outcome to every apparent deliberation, if the future holds no alternative pathways of possibility, then moral responsibility for guiding events is ruled out. Each actor is then merely living out his or her time line, experiencing what was in store for him or her at those later clock times. Such an actor, then, is not really a moral agent subject to moral praise or blame, since such a one is not cocreating a future with other agents responsible for actualizing events that could have been avoided under exactly the same circumstances but by different decisions.

This, for the question of goodness, is the primary inadequacy of the traditional mathematical approach to time that I criticized earlier from an epistemological point of view. On that theory of time, we recall, all events have a place in a continuous four-dimensional plenum, existing tenselessly at various clock times. The difference between future and past is not real but only mind-dependent. But if this view is taken not merely as a grand abstraction based on the formalisms used by modern physics but also as an account of the universe in which we live, then the future must be pictured as no less complete and determinate—and no less intractable to alteration—than the past. There is little wonder that such a conception has been called, polemically, but with some justice, the "block universe." Our future, could we but know it, is (in a tenseless sense of "is") inescapably part of the universe. Our efforts contribute to our future, of course, and so a shallow definition of fatalism in which events are fated despite our efforts is not entailed; but in a larger sense, our very efforts themselves are fated to occur just where and when they do occur. No one is responsible; no one could have done otherwise; the illusion of personal responsibility is based on the illusion of temporal becoming. Einstein and others have found this vision deeply comforting, but since it is a comfort that requires (if pushed to its logical conclusion, as Einstein did not consistently do) abandoning altogether the category of moral goodness, it is a price too high to pay even for relief from guilt and stress.

If we are prepared to acknowledge, however, that the formalisms of modern physics do not force us to abandon as data for our thinking the concreteness of lived experience, including moral experience and the intuitions of responsible agency, we need not adopt any such theory from mathematical physics as adequate or true. Since our minds need not be coerced by such abstractions, what might we consider more positively to be the ultimate moral significance of genuine temporal becoming?

It is intriguing sometimes to pose this question dramatically, in terms of what ingredients we might need if we were inventing a universe—especially a universe in which genuine personal growth in moral goodness is prized and in which a fundamental goal is the nurturing of mature moral community through creative mutuality among genuinely responsible agents. Given such goals, one necessary condition would be a medium for making decisions and making mistakes, for learning from those mistakes and for making more decisions and more mistakes and continuing to learn, both from personal trials and from the reports of others. Regarding others, it would be necessary in such a universe to provide a context in which growing personal agents could interact, responding one to another in creative ways step by step, as in a dialogue. Such dialogues would contain overtures, misunderstandings, corrections, risks, discoveries, regrets, joys, losses, and gains. Time as the medium for mutual encounter and personal growth is the secret of the possibility of such a universe.

Space, too, can be appreciated for its ultimate moral significance in providing the medium for the simultaneous otherness that growing, responsible persons need for self-identity. This constitutes the moral significance of the separative function of space. On its prehensive function, however, space also holds together in essential relatedness groupings of differentiated but mutually interacting agents as they deepen in relationship and participation with one another. Moral selfhood is never possible without some distinguishable center of responsibility; but full moral selfhood is never possible without essential internal relatedness to other selves. Ideal moral community is neither homogenized nor monadic.

Space-time, then, is wholly vital for the possibility of morally responsible growth in mutuality among persons. If we, in pondering what would be needed for inventing a universe aimed at nurturing such goodness, have realized that the medium of time is of the essence for our plan, perhaps God made the same discovery eons ago.

29.3. Time and the Sacred

Having now invoked the name of God, it seems appropriate—though daunting—to consider how time relates to whatever is taken as most high and most real, i.e., to whatever is our most intensive and comprehensive focus of valuation. This, as I define it, is the domain of the sacred.

The Christian tradition offers a conception of the Godhead as the tripartite paradigm of loving fellowship, of mutuality manifested in a unity of distinguishable but essentially related persons. This, quite apart from the history of creeds and councils, and even apart from the numerology of "threeness," is a beautiful affirmation of the ultimacy of the sort of goodness considered in the previous section. God is love. God is differentiated unity. God is fellowship. The sacred is love; the sacred is differentiated unity; the sacred is fellowship.

To take this vision of the sacred seriously, however, demands that we consider whether God could be loving fellowship apart from what we have found to be the necessary condition for creative interaction among mutually related persons. It seems, that is, that the nature of the Godhead, so understood, demands a form of time as medium for the loving interaction, for the creative dialogue, that underlies the ideal of fellowship. God's "discovery" of the essential role of temporal becoming was self-discovery. Such temporal becoming need not be imagined to be just like our experienced time; on the contrary, ours would need to be a thin, pedagogically oriented analogy drawn from the fullness of loving interaction that should be conceived at the heart of the sacred. God did not invent time, then; God's nature must contain time.

Thus we see that an adequate concept of God can hardly omit a fundamental sort of temporal becoming if becoming is itself a necessary ingredient in the highest value we know. Just as the concept of God as fellowship heals the ancient tension between the One and the Many, so this identification of the sacred also integrates the rift between Being and Becoming.

To this end, at least a few remarks about Being are in order. The thrust of all the foregoing has been to emphasize time and its significance, but Becoming alone can hardly have the final word if the full intuitions of humanity are to be given their due. I have mentioned earlier Plato's sense of the transcendent value of the eternal and unchanging, and we saw then that at least some aesthetic experience draws its power from the contemplation of permanent perfections in formal relationships. In this connection, we need to acknowledge the degree to which time, with its ever-gnawing tooth, is perceived as the enemy of value as well as the medium and context for its creation. One form of mystical experience, too, stresses the overcoming of time in the presence of the final sacred reality.

Those conceptual formalisms through which we attempt to represent the sacred, therefore, will need to attend to and incorporate elements from these data of widespread aesthetic, metaphysical, moral, and religious intuitions. Any "process theology" that omits the pole of transtemporality does so at the peril of its claim to adequacy. Fortunately, Whitehead himself never identified process with all of reality. There are many resources in his comprehensive conceptual framework for the needed recognitions of the eternal as well as the enduring.

Perhaps his generous formalisms may even provide the basis for integrating the contemplative and cosmic intuitions of Asian religions with the moral dynamism of the personalistic West.

One caution remains, however, as we pursue ever-increasing levels of coherence and adequacy for our theories about the sacred. I began by noting the historically conditioned character of all the formalisms of scientific thought. We should now end by acknowledging that such temporal conditions apply no less to all our philosophical and theological formulae as well. Every theory is a fabric of abstractions, and every abstraction must finally be inadequate to the infinitely rich concreteness of life itself. Just as our scientists must ever revise in the light of the stubborn weight of unformulated experience (despite the reciprocal pressure of theory to shape the ways that we experience), so must our philosophers and theologians revise not only in terms of the best science of the day but also in terms of the formulations of poets and moralists, educators and politicians, and of the unarticulated drift of human experience. It is a profoundly difficult task, but none is more vital for our fundamental orientation to existence itself. Perhaps the task is made slightly less pretentious, though no easier, by the recognition that for human beings under the rule of time, neither our little stabs toward goodness nor our approximations of truth are sacred in themselves.

Notes on Contributors

Andrew Bjelland teaches in the Department of Philosophy, Seattle University, Seattle, Washington 98122.

Ian G. Barbour teaches in the Departments of Religion and of Physics, Carleton College, Northfield, Minnesota 55057.

David Bohm has recently retired from the Department of Physics, Birkbeck College, University of London, London WC1E 7HX, England.

John B. Cobb, Jr., teaches at the School of Theology at Claremont and Claremont Graduate School, and is Director of the Center for Process Studies, 1325 N. College, Claremont, California 91711.

Milič Čapek has retired from the Department of Philosophy at Boston University, and is now adjunct professor at the Center for Science and Culture, 28 West Delaware Avenue, Newark, Delaware 19711.

Joseph E. Earley teaches in the Department of Chemistry, Georgetown University, Washington, D.C. 20057.

Tim Eastman teaches in the Department of Physics, University of Iowa, Iowa City, Iowa 52242.

Frederick Ferré is chairman of the Department of Philosophy, University of Georgia, Athens, Georgia 30602.

Crockett L. Grabbe is Associate Research Scientist in the Department of Physics and Astronomy, University of Iowa, Iowa City, Iowa 52242.

David Ray Griffin teaches at the School of Theology at Claremont and Claremont Graduate School, and is Executive Director of the Center for Process Studies, 1325 N. College, Claremont, California 91711.

Pete A.Y. Gunter teaches in the Department of Philosophy, North Texas State University, Denton, Texas 76203.

Patrick Hurley teaches in the Department of Philosophy, University of San Diego, Alcalá Park, San Diego, California 92110.

William B. Jones teaches in the Department of Philosophy, Old Dominion University, Norfolk, Virginia 23508.

Peter Miller teaches in the Department of Philosophy, University of Winnipeg, Winnipeg, Manitoba, Canada R3B 2E9.

Ilya Prigogine is a member of the Faculté des Sciences, Université Libre de Bruxelles, Campus Plaine, C.P. 231, 1050 Bruxelles, Belgium, and also of the Ilya Prigogine Center for Studies of Statistical Mechanics, University of Texas, Austin, Texas 78712.

Steven M. Rosen teaches in the fields of psychology and the philosophy and poetics of science at the College of Staten Island of the City University of New York, 715 Ocean Terrace, Staten Island, New York 10301.

Robert John Russell teaches at the Graduate Theological Union and is Director of the Center for Theology and the Natural Sciences, Graduate Theological Union, 2465 Leconte Ave., Berkeley, California 94709.

Henry P. Stapp is a member of the Theoretical Physics Group at the Lawrence Berkeley Laboratory, University of California, Berkeley, California 94720.

Name Index